BIOLOGICAL AND MEDICAL PHYSICS SERIES

Springer
Berlin
Heidelberg
New York
Barcelona
Hong Kong
London
Milan
Paris
Tokyo

Physics and Astronomy ONLINE LIBRARY

http://www.springer.de/phys/

BIOLOGICAL AND MEDICAL PHYSICS SERIES

The field of biological and medical physics is a broad, multidisciplinary, and dynamic one, touching on many areas of research in physics, biology, chemistry, and medicine. The Biological and Medical Physics Series is intended to be comprehensive, covering a broad range of topics important to the study of biological and medical physics. Its goal is to provide scientists, medical doctors and engineers with text books, monographs and reference books to address the growing need for information.

Editor-in-Chief:
Elias Greenbaum, Oak Ridge National Laboratory, Oak Ridge, Tennessee, USA

Editorial Board:
Masuo Aizawa, Department of Bioengineering, Tokyo Institute of Technology, Yokohama, Japan

Norma Allewell, Department of Biochemistry, University of Minnesota, St. Paul, Minnesota, USA

Olaf S. Andersen, Department of Physiology, Biophysics & Molecular Medicine, Cornell University, New York, USA

Robert H. Austin, Department of Physics, Princeton University, Princeton, New Jersey, USA

James Barber, Department of Biochemistry, Imperial College of Science, Technology and Medicine, London, England

Howard C. Berg, Department of Molecular and Cellular Biology, Harvard University, Cambridge, Massachusetts, USA

Victor Bloomfield, Department of Biochemistry, University of Minnesota, St. Paul, Minnesota, USA

Robert Callender, Department of Biochemistry, Albert Einstein College of Medicine, Bronx, New York, USA

Britton Chance, Department of Biochemistry/ Biophysics, University of Pennsylvania, Philadelphia, Pennsylvania, USA

Steven Chu, Department of Physics, Stanford University, Stanford, California, USA

Louis J. DeFelice, Department of Pharmacology, Vanderbilt University, Nashville, Tennessee, USA

Johann Deisenhofer, Howard Hughes Medical Institute, The University of Texas, Dallas, Texas, USA

George Feher, Department of Physics, University of California, San Diego, La Jolla, California, USA

Hans Frauenfelder, CNLS, MS B258, Los Alamos National Laboratory, Los Alamos, New Mexico, USA

Ivar Giaever, Rensselaer Polytechnic Institute, Troy, New York, USA

Sol M. Gruner, Department of Physics, Princeton University, Princeton, New Jersey, USA

Judith Herzfeld, Department of Chemistry, Brandeis University, Waltham, Massachusetts, USA

Pierre Joliot, Institute de Biologie Physico-Chimique, Fondation Edmond de Rothschild, Paris, France

Lajos Keszthelyi, Institute of Biophysics, Hungarian Academy of Sciences, Szeged, Hungary

Robert S. Knox, Department of Physics and Astronomy, University of Rochester, Rochester, New York, USA

Aaron Lewis, Department of Applied Physics, Hebrew University, Jerusalem, Israel

Stuart M. Lindsay, Department of Physics and Astronomy, Arizona State University, Tempe, Arizona, USA

David Mauzerall, Rockefeller University, New York, New York, USA

Eugenie V. Mielczarek, Department of Physics and Astronomy, George Mason University, Fairfax, Virginia, USA

Peter B. Moore, Department of Chemistry, Yale University, New Haven, Connecticut, USA

V. Adrian Parsegian, Physical Science Laboratory, National Institutes of Health, Bethesda, Maryland, USA

Linda S. Powers, NCDMF: Electrical Engineering, Utah State University, Logan, Utah, USA

Earl W. Prohofsky, Department of Physics, Purdue University, West Lafayette, Indiana, USA

Andrew Rubin, Department of Biophysics, Moscow State University, Moscow, Russia

Michael Seibert, National Renewable Energy Laboratory, Golden, Colorado, USA

David Thomas, Department of Biochemistry, University of Minnesota Medical School, Minneapolis, Minnesota, USA

Samuel J. Williamson, Department of Physics, New York University, New York, New York, USA

Nikolai L. Vekshin

Photonics of Biopolymers

With 94 Figures

 Springer

Dr. Nikolai L. Vekshin
Russian Academy of Science
Institute of Cell Biophysics
142290 Moscow Region, Pushchino
Russia

Cataloging-in-Publication data applied for
Die Deutsche Bibliothek - CIP-Einheitsaufnahme
Vekshin, Nikolai L.:
Photonics of biopolymers / Nikolai L. Vekshin
Berlin; Heidelberg; NewYork; Barcelona; Hong Kong; London; Milan; Paris; Tokyo: Springer, 2002
 (Biological and medical physics series)
 (Physics and astronomy online library)
  ISBN 3-540-43817-3

An earlier edition of this book was published in 1999 by Moscow University Press

ISSN 1618-7210
ISBN 3-540-43817-3 Springer-Verlag Berlin Heidelberg New York

This work is subject to copyright. All rights are reserved, whether the whole or part of the material is concerned, specifically the rights of translation, reprinting, reuse of illustrations, recitation, broadcasting, reproduction on microfilm or in other ways, and storage in data banks. Duplication of this publication or parts thereof is permitted only under the provisions of the German Copyright Law of September 9, 1965, in its current version, and permission for use must always be obtained from Springer-Verlag. Violations are liable for prosecution act under German Copyright Law.

Springer-Verlag Berlin Heidelberg New York
a member of BertelsmannSpringer Science + Business Media GmbH

http://www.springer.de

© Springer-Verlag Berlin Heidelberg 2002
Printed in Germany

The use of general descriptive names, registered names, trademarks, etc. in this publication does not imply, even in the absence of a specific statement, that such names are exempt from the relevant protective laws and regulations and therefore free for general use.

Typesetting: Digital data supplied by author
Cover concept: eStudio Calamar Steinen using a background picture from The Protein Databank (1 Kzu). Courtesy of Dr. Antoine M. van Oijen, Department of Molecular Physics, Huygens Laboratory, Leiden Unversity, The Netherlands. Reprinted with permission from Science 285 (1999) 400-402 („Unraveling the Electronic Structure of Individual Photosynthetic Pigment-Protein Complexes", by A.M. van Oijen et al.) Copyright 1999, American Association for the Advancement of Science.
Cover production: *design & production* GmbH, Heidelberg

Printed on acid-free paper   SPIN 10852124   57/3020/Rw   5 4 3 2 1 0

Preface

Absorption and emission of the UV/visible light are widely applied phenomena in biophysical and biochemical investigations. A number methods for spectroscopic studies of biopolymers and membranes are based on a solid physical background. It is therefore important to clarify the physical peculiarities of the energy transformation processes in photoexcited molecular-biological structures. This book discusses the features of such processes as light absorption, fluorescence emission, energy and electron transfer, exciplex and excimer formation, relaxations and chemical reactions in solutions of proteins, nucleic acids, membranes and model systems. The most interesting experimental data are subjected to analysis and some typical examples are given. A considerable part of the book is based on my fundamental work and recent experience. I hope that this book will be useful for everyone who applies optical spectroscopy to investigate the structural and functional properties of biological macromolecular systems.

This book is a second edition (with a few corrections and additions) of *Photonics of Biopolymers*, which was published in English by the Moscow State University Press in 1999. I am very grateful to three persons who contributed the greatest editoral help in the preparation of the initial manuscript: Prof. S. Schulman (University of Florida, USA), Prof. O. Wolfbeis (Regensburg University, Germany), and Dr. D. Toptygin (Johns Hopkins University, Baltimore, USA).

July 2002   Nikolai Vekshin

Contents

1 Introduction to Photonics 1
   1.1 Light Absorption 3
   1.2 Vibrational Relaxation and Internal Conversion 3
   1.3 Fluorescence 3
   1.4 Intersystem Crossing and Phosphorescence 4
   1.5 Energy Transfer 5
   1.6 Excimer and Exciplex Formation 5
   1.7 Photochemical Reaction and Electron Transfer 5
   1.8 Photoinduced Conformational Changes 5

2 Light Absorption in Ordered Structures 7
   2.1 Probability of Absorption 7
   2.2 Lambert–Beer Law 9
   2.3 Biological Chromophores 10
   2.4 Absorbance in Scattering Media 11
   2.5 Hypochromism 11
   2.6 Molecular Interactions 12
   2.7 Sieve Effect 13
   2.8 Light-Scattering Model 14
   2.9 Light-Dispersion Model 15
   2.10 Stacking Model 15
   2.11 Conclusion 16

3 Screening Hypochromism 18
   3.1 Screening model 19
   3.2 Nucleotides, Oligonucleotides and DNA 23
   3.3 Tyrosine and Tryptophan in Aggregates 27
   3.4 Tyrosine and Tryptophan in Proteins 28
   3.5 Hemoglobin in Erythrocytes 30
   3.6 Chloroplasts and Visual Rods 31
   3.7 Clusters of Aromatic Hydrocarbons 32
   3.8 Conclusion 35

4 Photometric Estimation of Protein Content in Biological Suspensions 36
   4.1 Colorimetry and Others Methods 36
   4.2 Combined UV-Spectrophotometric Method 36

VIII Contents

4.3 Comparison of the Methods 38
4.4 Conclusion 40

5 Screening and Reabsorption of Light 41
5.1 Inner Filter Effect 41
5.2 Microscreening and Microreabsorption 42
5.3 Erythrocytes and Other Cells 44
5.4 Volume Reabsorption of Donor Luminescence 45
5.5 Trypaflavine Activates Rhodamine Fluorescence 47
5.6 Conclusion 50

6 Multipass Cuvettes for Luminescence Spectroscopy 51
6.1 Mirror and Total Internal Reflection Cuvettes 51
6.2 Multipass Cuvettes in Steady-State Measurements 53
6.3 Multipass Cuvettes in Lifetime Measurements 54
6.4 Other Applications 55
6.5 Conclusion 55

7 Division of Tyrosine and Tryptophan Fluorescence Components 56
7.1 Generally Accepted Approaches 56
7.2 Synchronous Scanning Method 58
7.3 Synchronous Spectra of Tryptophan and Tyrosine 58
7.4 Synchronous Spectra of Different Tryptophan Residues 60
7.5 Conclusion 61

8 Spectral Heterogeneity of Tryptophan Emission 62
8.1 Variation of Fluorescence Polarization along Tryptophan Emission
Spectrum 62
8.2 Fluorescence Lifetime Variations 68
8.3 Photoinduced Conformational Mobility of Proteins 70
8.4. Phosphorescence of Proteins 72
8.5 Conclusion 73

9 Discrete Emission States in Photoexcited Tryptophan Complexes ...74
9.1 Time-Resolved Spectroscopy of Tryptophan Fluorescence 75
9.2 TRP and NATA in Water 78
9.3 TRP, NATA and Indole in Ethanol 80
9.4 TRP and NATA in Glycerol 81
9.5 Dipeptides 83
9.6 Exciplexes in Proteins 84
9.7 Conclusion 88

10 Mechanisms of Exciplex Formation 90
10.1 Generally Accepted Models 90
10.2 Fractional Energy Transfer in Exciplexes 92
10.3 Exciplex of Aromatic Hydrocarbons 94

10.4 Excimers 96
10.5 Pyrene-Indole Exciplex 98
10.6 Conclusion 100

11 Mechanisms of Energy Transfer 101
11.1 Inductive-Resonance Model 102
11.2 Energy Transfer in Molecular Structures 105
11.3 Hot Migration 108
11.4 Conclusion 110

12 Energy Transfer in Nucleic Acids 111
12.1 Migration Between Nucleotides 111
12.2 Migration Along DNA 113
12.3 Quantum Yield of Energy Transfer to Dyes 113
12.4 Polyadenilic Acid Labeled by Ethenoadenine 115
12.5 DNA with Intercalated Dyes 115
12.6 Fluorescent Probes and Labels on DNA 119
12.7 Conclusion 121

13 Energy Transfer in Native Proteins 122
13.1 Tyrosine-Tryptophan Pair 122
13.2 Migration between Tryptophan Residues 126
13.3 Tryptophan-NADH Pair in Alcohol Dehydrogenase 127
13.4 Tryptophan-Heme Pair in Myoglobin 132
13.5 Tryptophan-Pyrene Pair 134
13.6 Quenching of Tryptophan Emission by Dyes 137
13.7 Conclusion 141

14 Energy Transfer in Biomembranes 142
14.1 Quenching of Tryptophan Fluorescence in Sarcoplasmic Reticulum by Probes 143
14.2 Quenching of Tryptophan Fluorescence by ANS 145
14.3 Quenching of Tryptophan Fluorescence by Pyrene 146
14.4 Tryptophan-NADH Pair in Mitochondria 147
14.5 Photosynthetic Reaction Centers 149
14.6 Conclusion 150

15 Fluorescence Probes 151
15.1 Widely Used Probes 151
15.2 Estimation of Sizes of Chaperones and their Complexes Using ANS 155
15.3 Fluorescent Studies of Na^+K^+-ATPase 158
15.4 Anthracene with Dimethylaminochalcone in Membranes 159
15.5 Diffusion of Probes 161
15.6 Fluorescence Pharmacology *in vitro* 162
15.7 Conclusion 163

16 Pyrene Monomers and Excimers in Membranes 165
  16.1 Viscosity Measurements 165
  16.2 Location and Diffusion of Pyrene 166
  16.3 Detection of Oxygen by Pyrene Emission 168
  16.4 Vibronic Peaks as Indicators of Membrane Polarity 170
  16.5 Conclusion 171

17 Photomodulation of Enzyme Activity 172
  17.1 Photoactivation of Enzymes 172
  17.2 Photodesorption 173
  17.3 Photochemical Processes in Alcohol Dehydrogenase 174
  17.4 Photolysis of Flavin in NADH Dehydrogenase 179
  17.5 Conclusion 183

18 Photoactivation of Animal Membranes and their Chromophores 184
  18.1 Photoinduced Membranes Activity 184
  18.2 Oxygen Uptake in Mitochondria under Photoexcitation 185
  18.3 Oxidation of NADH by Triplet Flavin and Singlet Oxygen 187
  18.4 Conclusion 192

19 Light-Dependent Phosphorylation in Mitochondria 193
  19.1 ATP Synthesis during Illumination 193
  19.2 Thermal Coupling between ATP Synthesis and Electron Transfer 197
  19.3 Conclusion 199

References 200

Index 228

1 Introduction to Photonics

Light absorption in biostructures results in energy transformation processes similar to those in molecular solutions, i.e. vibrational relaxation, internal conversion, intersystem crossing to a triplet state, fluorescence and phosphorescence emission, electronic excitation energy transfer, excimer and exciplex formation, conformational changes, electron transfer, and chemical reactions. Some of these processes are depicted in Fig. 1.1. The study of such processes is the subject of photonics. This book analyzes the concepts of energy transformation in photoexcited proteins, nucleic acids, membranes and respective model systems and discusses experimental data on these subjects.

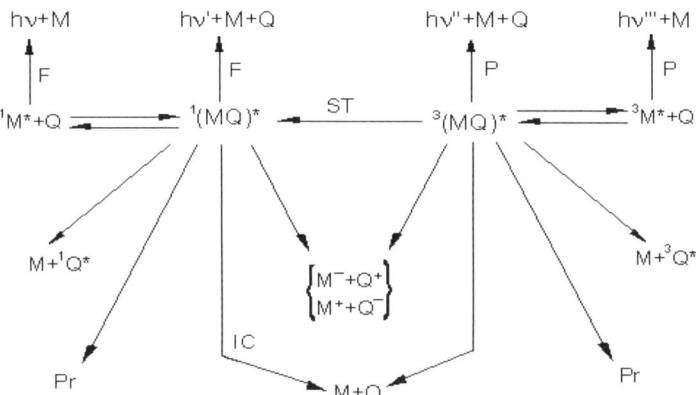

Fig. 1.1. Energy transformation pathways in photoexcited chromophore systems [53], with modifications. M is a monomer molecule in the ground state, Q is a quencher or an energy acceptor, Pr is a product of a photochemical reaction, IC stands for internal conversion, ST stands for singlet-triplet intersystem crossing, F stands for fluorescence, P stands for phosphorescence

The need for developing a specific *photonics of biopolymers* results from the following facts:
(a) usually the notions used in photonics of molecular solutions are applied to biostructures [1]. This is often justified. However, heterogeneity, anisotropy, orderliness, relatively high rigidity, and some other properties of biomacromolecules and membranes may result in uncommon relations between different processes of photoexcitation energy transformation and in quite unexpected spec-

tral effects. The study of light absorption and energy transformation in molecular biosystems is essential for understanding basic physical mechanisms of these processes and the consequences they have in heterogeneous, anisotropic, ordered, or rigid structures. A considerable part of the monograph deals with this subject.

(b)   Some processes of photoexcitation energy transformation can be used as the physical bases for studying the structure and function of membranes and biopolymers [2]. The methods are based on interaction of light with intrinsic chromophores occurring naturally in these objects or with extrinsic fluorescent labels. For instance, millisecond kinetics of enzyme folding and refolding is studied using tryptophan fluorescence [3–5]. Fluorescence spectroscopy, the most powerful method in molecular photonics [2, 6–12], provides valuable information on spatial organization and functioning of biostructures [13–27]. New developments in practice of molecular luminescence for chemical and biochemical analysis in UV– visible and near-infrared regions were briefly reviewed in [28].

(c)   In many biological systems, energy transformation is the basis for specific biological functions. Examples of such processes include bioluminescence [29–33], production of vitamin D and pigments in skin [34], transformation of light quanta into electric signals in rods and cones in the retina [34], and most importantly – bacterial and plant photosynthesis [35–44].

(d)   In recent years, lamp and laser stimulation has been widely used in medicine and cellular biology.

(e)   Spectral properties of biomolecules can be used to design biosensors [45–48].

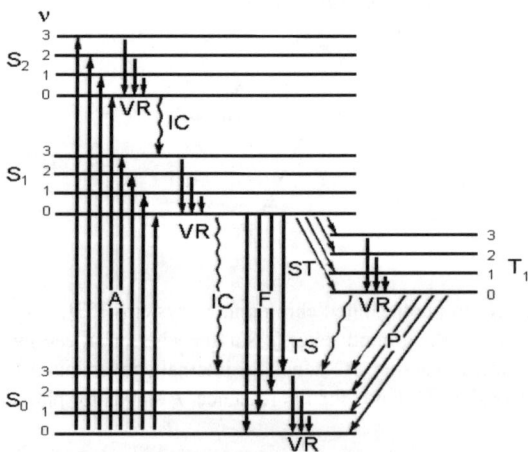

Fig. 1.2. The Jablonski diagram: photophysical processes occurring in a conjugated molecule on interaction with light (for abbreviations see the legend of Fig. 1.1 and in the text)

Some of the energy pathways following photoexcitation of a chromophore are shown in Fig. 1.2 (the diagram is from [11] with insignificant modification). Electronic transition from the lowest vibrational level ($v = 0$) of the ground state (S_0) to the vibrational levels ($v = 0, 1, 2, 3$) of the excited singlet states (S_1 or S_2)

takes place with some probability when a chromophore molecule interacts with a photon. Then, intramolecular electron-vibrational interactions lead to rapid vibrational relaxation (VR) or internal conversion (IC). If VR takes place, the molecule finds itself in the low vibrational levels of S_1 state. Competing for deactivation of S_1 are IC, ST and fluorescence. IC or fluorescence transfers the molecule to the S_0 state. ST transition is intersystem crossing from the singlet excited state to a triplet one. Phosphorescence and triplet-singlet intersystem crossing (TS) deactivate the molecule to S_0.

1.1 Light Absorption

When a molecule is illuminated by ultraviolet (UV) or visible light, its chromophore can absorb a photon. The probability (P) of absorption depends, in particular, on the energy of the photon (on the wavelength or frequency of light) and on the energy levels of the chromophore. P does not depend on the intensity of light as long as the intensity is low (i.e., a molecule interacts with a single photon). The period of an absorption transition is no more than a few femtoseconds.

1.2 Vibrational Relaxation and Internal Conversion

For conjugated molecules, vibronical relaxation (VR) and internal conversion (IC) occur within about 0.1–1.0 ps. The VR and IC transition results from intrinsic molecular rotations and vibrations. VR energy losses range from 100 to 5000 cm^{-1}, which corresponds to transitions in the infrared spectrum. Infrared emission of photoexcited proteins were detected in [49].

1.3 Fluorescence

Radiative transitions from the S_1 to the S_0 state occur in a time frame of 0.01–100 ns. Such luminescence is called fluorescence. Since the rate of light emission is much slower than the rate of VR, the fluorescence spectrum is shifted to the "red" as compared to the absorption spectrum. In solutions, interactions between solute and solvent also contribute to the red shift in the emission spectrum.

The probability of fluorescence (quantum yield) can be expressed in terms of rate constants of competing processes:

$$Q_f = K_f / [(K_f + K_{ic} + K_{st}] \tag{1.1}$$

where K_f is the rate constant of fluorescence, K_{ic} and K_{st} are the rate constants of IC and ST processes. When a given excited molecule interacts with a non-excited one, other rate constants (rates of energy transfer, electron transfer, excimer formation, deactivation, etc.) must be introduced in the denominator of (1.1). The greater are the rates of the processes competing with radiative transitions from S_1, the lower is the value of Q_f.

The rate constant of the decay equals the inverse of the lifetime, therefore:

$$Q_f = \tau/\tau' \tag{1.2}$$

where τ is the measured fluorescence lifetime and τ' is the natural excited-state lifetime [50].

The fluorescence lifetime is described by an exponential law:

$$N(t) = N_0 e^{-t/\tau} \tag{1.3}$$

where N_0 is the initial population of excited chromophores, $N(t)$ is the number of excited chromophores at time t after an excitation impulse.

Q_f can be expressed in terms of the intensities of the incident and the transmitted light flux (I' and I'', respectively) and the intensity of fluorescence (F):

$$Q_f = F/(I'-I'') \tag{1.4}$$

Using (1.4) and the well-known Lambert-Beer law (2.4), one can obtain an approximate equation, which is valid for small optical densities (absorbance, $A<0.02$):

$$Q_f = F/2.3 I' \varepsilon [c] L \tag{1.5}$$

where ε is the extinction coefficient, $[c]$ is the concentration and L is the optical path in the sample.

In fluorescence measurements it is important to eliminate the possibility of instrumental artifacts, which may create illusions of false effects or, conversely, mask true effects. The most frequent artifacts encountered in fluorescence instrumentation are the following: (a) penetration of excitation light into the registration channel; (b) Raman scattering; (c) the presence of the second or third diffraction orders in grating monochromators; (d) photochemical transformations in the sample; (e) variation of photomultiplier sensitivity with the wavelength; (f) the effect of monochromators on the polarization of transmitted light. For instance, when a system is excited with an unpolarized light and the fluorescence is detected without polarizer, the emission intensity from viscous or frozen samples may be changed due to the monochromator polarization effect up to 20%. Using the detection at 55° or two polarizers positioned at the same "magic" angle, the artifact is totally abolished [51].

1.4 Intersystem Crossing and Phosphorescence

Intersystem crossing results in a change in the spin of an electron. The energies of ST-transitions are sufficient for changing the spin. The radiative transition from T_1 to S_0 is a spin-forbidden process; and lasts rather long (0.1 ms–10 s). This kind of luminescence is called phosphorescence. In solutions, phosphorescence is strongly quenched by the solvent, oxygen, etc. Phosphorescence can be detected in glasses at liquid nitrogen temperature or sometimes at room temperature provided the chromophore does not experience strong interactions with its microenvironment. Equations similar to (1.1–1.5) can be applied to phosphorescence.

1.5 Energy Transfer

When an electronically excited molecule (donor) interacts with a nonexcited molecule (acceptor), nonradiative transfer of electronic energy from the donor to the acceptor may take place. If the acceptor is fluorescent or phosphorescent, sensitized luminescence of the acceptor will be registered.

This energy transfer can be represented as follows: **D*+A→D+A*,** where **D*** is the donor molecule in the electronically excited state and **A** is the acceptor molecule in the ground state. Energy transfer can be detected by the decrease in the quantum yield or in the lifetime of the donor and by the appearance of sensitized luminescence of the acceptor.

1.6 Excimer and Exciplex Formation

When a photoexcited molecule collides with a ground state molecule, an excited state complex can be formed which are referred to as "exciplexes" [52–55]. A wide structureless exciplex emission band shifted to the "red" is observed. The emission from the exciplex does not originate from the quencher molecule, but is rather produced by the complex as a whole. An exciplex formed by two identical molecules (a dimer in the excited state, which is formed from one excited and one unexcited molecule) is called an "excimer" [52–55]. If formed by two different species, the complex is referred to as a "heteroexcimer" or "exciplex".

1.7 Photochemical Reaction and Electron Transfer

When a photoexcited donor molecule collides with a quencher, an electron can be transferred from the donor to a free orbital on the quencher molecule. Usually photoexcitation results in accelerated redox reactions or in other chemical transformations. Photochemical reactions, their kinetics and mechanism have been reviewed [6, 8, 53, 56, 57].

1.8 Photoinduced Conformational Changes

Structurally flexible molecules may undergo conformational changes on photoexcitation. These changes result from vibrations and rotations, which accompany nonradiative VR and IC transitions. If the molecule is rigid, then only relaxational movement of the molecule in its microenvironment can be observed. Two driving forces govern relaxation dynamics (a) in the excited state the dipole moment is increased, which results in a different energy of interaction with solvent shell, and (b) VR, IC, ST or TS processes which are accompanied by local "heating" of the chromophore and its microenvironment.

2 Light Absorption in Ordered Structures

2.1 Probability of Absorption

The probability of absorption of a photon by a molecule (at a given wavelength) can be represented as the ratio of two areas [6, 58, 59]:

$$P = \delta/X \quad (2.1)$$

where δ is the absorption cross-section, i. e. "the area impermeable for a photon" [6], and X is the cross-sectional area of the molecule; δ does not exceed X. $\delta(\text{Å}^2)$ should be converted to the molecular extinction coefficient $E(\text{Å}^2)$ The conversion from δ to E requires to multiply E by 2.3, i.e., $\delta = 2.3E$. The value 2.3 arises from replacing natural logarithms by decimal ones [58, 59]. $E(\text{Å}^2)$ can be converted directly to the molar extinction coefficient ε (M^{-1}cm^{-1}). There is a direct conversion of the dimension Å^2 into M^{-1}cm^{-1}, for instance, 1 Å$^2 = 6\times10^7$ cm^3/6×10^{23} cm = 6×10^4 litre/mol cm = 60000 M^{-1}cm^{-1}. To take into account the dependence of absorbing ability of the chromophore on its orientation [6, 60–62], the orientation factor (q) should be included in the equation. Finally, instead of the ambiguous parameter X, which is not strictly defined [6], it is necessary to use the more exact parameter - area of density of the electrons which is responsible for optical transition, i. e. the cross-section of the valence electrons (S). Now we can rewrite (2.1) as follows [63]:

$$P = 2.3Eq/S \quad (2.2)$$

S is described as the projection of maximum cross-section of electron density responsible for a given optical transition. S can be predicted by quantum-chemical methods [64] or obtained from low-temperature x-ray analysis [65]. The typical values of the area for single bonds between carbon, hydrogen, oxygen or nitrogen atoms and for lone electron pairs on oxygen or nitrogen atoms are 0.32–0.91 Å^2 [64]. For double and conjugate bonds between carbon atoms, S is about 0.9–1.5 Å^2 per bond. A value of S for a molecule is obtained by summing all the elementary electron cross-sectional areas (cross-sections of electron density) participating in the absorption of a photon by the chromophore.

Absorption of photons is most efficient when the oscillations of the electric vector of the light wave (which is perpendicular to the direction of propagation) coincide with the direction of electronic transition, which lies in the plane of the

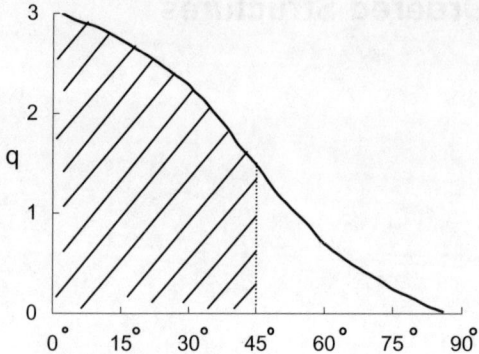

Fig 2.1. The dependence of the orientation factor for a linear oscillator on the spherical angles between the transition dipole and electric vector. The marked area corresponds to the one half of molecules that contribute to ~80% of the absorption

chromophore. In this case δ is three times larger than in the case of a random (chaotic) orientation [1, 68, 73]. Parallel and random orientations correspond to $q=3$ and $q=1$ respectively. If the electric vector of the light wave is perpendicular to the chromophore plane, q approaches 0. The dependence of absorbtion on the angles between light flow and chromophore orientations in solution was described in [66]. The dependence of q on the spherical angles is given in Fig. 2.1. It appears that in solution approximately 80% of absorption results from one half of all molecules (these molecules have an average q of about 2) and approximately 20% comes from the other half (with an average q of about 0.5).

The following example [67] shows the calculation of P for anthracene using (2.1). "The extinction coefficients of anthracene are 160000 and 6300 $M^{-1}cm^{-1}$ at 253 and 375 nm, respectively. These values correspond to cross sections of 6.1 and 0.24 Å^2, respectively. Assuming the molecular cross section of anthracene to be 12 Å^2, we see anthracene absorbs about 50% of the photons it encounters at 253 nm and 2% of the photons at 375 nm". We have calculated the probabilities of absorption for a number of biological chromophores (Table 2.1) using an approach similar to the one described above. Our approach, however, was more accurate, instead of the rough value of "the area of the molecule" (X) we used S (2.2)

Because of the fact that for many aromatic and heteroaromatic molecules the maps of electron density distribution of the optical electron do not yet exist, we will rely on maps for separate bonds and fragments from [64, 65]. The cross-sectional area of electron density depends on orientation.

In diluted solutions of single identical molecules the probability of absorption of a photon is the same for each chromophore:

$$P=P'=P''... \qquad (2.3)$$

In this case the intensity of the light beam in solution decreases according to thenwell-known Lambert-Beer law (2.4). In contrast, in aggregates and macromo-

Table 2.1. Extinction coefficients, absorption sections, cross-sectional areas of electron density, and absorption probabilities for some chromophores in solution ($q = 1$) [63]

Substance	λ_m [nm]	ε [M^{-1}cm^{-1}]	E [Å^2]	δ [Å^2]	S [Å^2]	P
Adenine	260	15000	0.25	0.575	7.1	0.081
Tyrosine	275	1270	0.021	0.048	3.1	0.016
Tryptophan	280	5600	0.093	0.214	4.3	0.05
Retinol	498	40000	0.7	1.61	5.4	0.3
Porphyrin	408	120000	2.0	4.6	14.5	0.317
Chlorophyll	430	120000	2.0	4.6	17.0	0.271
Anthracene	253	200000	3.3	7.59	8.0	0.95

lecules, where inter-chromophore distances are much smaller than the wavelength of light, competition for a photon is created: the probability of absorption of a photon by each chromophore depends on the probability of absorption by others.

2.2 Lambert–Beer Law

In a solution of molecules absorbing light, the intensity of the light beam decreases exponentially. This can be described by the Lambert–Beer law:

$$A = \lg(I'/I'') = \varepsilon[c]L \qquad (2.4)$$

where A is the absorbance, I' is the intensity of incident light, I'' is the intensity of transmitted light, ε is the extinction coefficient, $[c]$ is the concentration, and L is the optical depth of the sample. The Lambert-Beer law is observed when chromophores do not shade each other from the light [6]. Since ε changes with the wavelength, one can record the absorption spectrum, i. e. the dependence of A on the wavelength.

If $[c]$ and L are expressed in moles/litre and cm, respectively, then ε has the dimension M^{-1}cm^{-1}. This dimension is equal to cm^2; i.e. it is a unit of area. In dilute solutions at common I' the value of ε is a molecular constant, which independent of I', I'', L and $[c]$.

If L in (2.4) is represented as $V/Q = L$, where V is the illuminated volume and Q is the illuminated area of the sample cell, and $[c]$ is represented as $N/V=[c]$, where N is the number of chromophore molecules in the volume, then one can write [68, 69].

$$A = EN/Q \qquad (2.5)$$

From such a representation of the Lambert-Beer law one may conclude that the absorbance depends on N and does not depend on the arrangement of chromophores in the illuminated area.

According to Lambert-Beer law, the absorbance is proportional to extinction coefficient, concentration of molecules, and optical path. In practice, the following effects may lead to deviations from the Lambert-Beer law: (a) light scattering (including Raman); (b) fluorescence and phosphorescence; (c) photochemical reactions and photobleaching; (d) aggregation; (e) changes in the orientation of anisotropic samples, (f) strong molecular interactions; (g) non-monochromaticity of the light; (h) hypochromic effect (including a mutual screening in chromophore stacks).

2.3 Biological Chromophores

Natural biological chromophores have strong absorption bands in the UV or visible regions. Many of them have sufficiently high quantum yields of emission. Spectroscopic properties of some biological chromophores are shown in Table 2.2.

Table 2.2 Spectroscopic properties of biological chromophores in aqueous solutions [18, 58, 60, 67, 70]

Chromophore	λ_m [nm]	ε [M^{-1}cm^{-1}]	F_m [nm]	Q_f	τ [ns]
Tryptophan	280	5600	348	0.14	3.0
Tyrosine	274	1400	303	0.20	3.6
Phenylalanine	257	200	282	0.04	6.4
Adenosine	260	13400	321	<0.0003	<0.02
NADH	340	6200	470	0.02	0.4
FMN	445	12500	530	-	4.6
FAD	450	11300	530	-	2.3
Protoporphyrin-X	621	-	623	0.10	12.4

λ_m is the long-wavelength absorption maximum, ε is the molar extinction coefficient, F_m is the fluorescence maximum, Q_f is the quantum yield of fluorescence, τ is the fluorescence lifetime.

For adenine, small values of Q_f and τ result from the high rate constant of internal conversion. For flavine adenine dinucleotide (FAD) and reduced nicotinamide adenine dinucleotide (NADH), small Q_f and τ result from quenching of the flavine chromophore of FAD and pyridine ring of NADH by the adenine chromophore in these molecules. Resonance energy transfer from flavin or pyridine to adenine is forbidden because of the energy barrier (the adenine absorption band is located at shorter wavelengths than the fluorescence of the flavin or pyridine chromophores). Therefore, the quenching is probably the result of deactivation by vibronic modes of the adenine chromophore (and also quenching by solvate molecules).

2.4 Absorbance in Scattering Media

Scattering of light beam by particles leads to an increase in the measurable absorbance of a suspension. In not too turbid media the absorbances and scattering are additive. Figure 2.2 shows the absorption spectra of adenosine monophosphate (AMP) in solution with and without glycogen and the spectrum, which is due to light scattering by glycogen.

The effect of light-scattering is almost completely eliminated when the cuvette with solution is placed inside a spherical detector or close to a half-sphere detector [59].

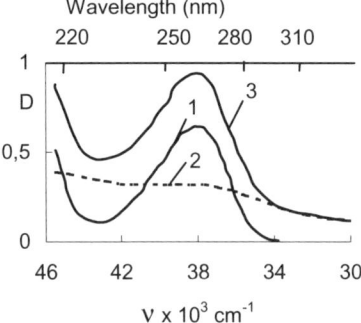

Fig. 2.2. Absorption spectrum of AMP (1), glycogen (2) and an AMP-glycogen mixture (3) recorded with 1-cm cuvettes

A simple method for subtraction of the scattering part from the total absorbance was suggested in [71]. The method consists in the fitting of a power law to an absorption – free segment of the total spectrum. The power parameter is related to the dimensions, refractive index and size polydispersity of the suspensions. The methodology was applied to a polyene sterol probe in lipid vesicles.

2.5 Hypochromism

One of the most interesting phenomena observed in stacked chromophores is the hypochromic effect, i.e. the decrease in the extinction coefficient of neighboring chromophores [72].

Experimentally the value (h) of hypochromism at a given wavelength is calculated from the equation:

$$h = 100\%(\varepsilon - \varepsilon^\wedge)/\varepsilon \qquad (2.6)$$

where ε is the extinction coefficient for solution of single chromophores, $\varepsilon^\wedge$ is the average extinction coefficient of a chromophore in aggregate or a macromolecule.

Hypochromism is observed in many molecular structures, for example dye aggregates, clusters of aromatic compounds, microcrystals, polymers, proteins, nucleic acids, chloroplasts, etc [6, 58, 60, 72, 73].

The accepted theory considers hypochromism as a result of weak dipole-dipole molecular interactions modified by the light wave [58, 74]. However, this theory was criticized in Bolton and Weiss [75] and Nesbet [76] and various modified and other models [75–81] were suggested. One of them interpreted hypochromism as a result of changes in the complex refractive index [78, 79, 82, 83]. Refractive index consists of two components: the natural component and imaginary one. The imaginary part relates to an absorption process [82]. The models demonstrate the importance not only of the refractive index but also of local light field [83, 84] and of dispersion of effective field of light wave [82, 85]. Some explanations of the phenomenon are given below.

2.6 Molecular Interactions

If chromophores are brought into spatial proximity, strong interactions occur between their electronic clouds, between charged or polar groups. The interactions result in redistribution of the electronic density. This is reflected in the absorption spectra: the intensity (oscillator strength) of one band decreases and the intensity of another increases or, in many cases, new absorption bands appear [6, 60, 86]. The total oscillator strength must be conserved. When the interaction energy is enough high (i.e., more than ~40 kJ/mol), the molecular interaction can lead to hypochromism and distortion of the absorption spectra. This has been shown for dimers of many dyes [1]. However, in the case of "weak" interactions (~4 kJ/mol or less) the UV and visible spectra do not change significantly. For example, it was noted [87] that the calculation of the hypochromism of purinophanes by the Pariser-Parr-Pople method under the assumption of the weak interactions provides no satisfactory agreement with experimental values. These values did not change with the distance in parallel-stacking pyrimidinopurinophanes [88]. The following equation from [89] was used in [87, 88] to fit the calculated values to the experimental ones:

$$h/100\% = C(1-3\cos^2\theta)r^{-3} \qquad (2.7)$$

where r is the radius vector or distance between chromophore centres, θ is the angle between the radius vector and the transition moment, and C is a coefficient. By fitting the value of the coefficient they were able to achieve some agreement with the experimental values. The physical sense of C remains un-clear, however.

Strong interactions of polar and charged dyes in dimers and polymers can give rise to "coulombic" hypochromism [6, 60, 90], which is usually accompanied by the distortion of the initial band, an increase in the absorption of another band, or the appearance of new bands. However, "both classical and quantum mechanical approaches yield the same results: a stacked arrangement of oscillators in a dimer makes the absorption band shift towards shorter wavelengths without changing the intensity" [91]. The coulombic interactions of chromophores in the ground state are actually very small in DNA, RNA, oligonucleotides, polypeptides, proteins and some polymers. The interactions in such systems are weak, and with the decrease in one absorption band the intensities of others do not change and new bands do not occur [58, 60, 92]. However, the IR absorption spectra are substantially affected.

2.7 Sieve Effect

In heterogeneous samples with particles of a large size, a part of the light flux can sieve between suspended particles ("sieve effect") [93–102]. Let us assume that molecules accumulate in a part of the illuminated volume for some reason. In this case the incident light can be represented by two luminous fluxes, one of which passes through suspended particles with an exponential decrease in its intensity, while the seconds flux sifts through the "empty" solvent without absorption. The absorption law will be nonexponential, and thus the absorbance of the suspended particles will be less than the one in the solution of single chromophores [93–98, 103]. This is true for very heterogeneous samples with particles larger than the wavelength. The equation obtained by Duysens [93] for suspensions of spherical particles is:

$$A_{sus}/A_{sol} = 3(1-T_p)/(2\pi r^2) \tag{2.8}$$

where A_{sus} is absorbance of suspension, A_{sol} is absorbance of a molecular solution, T_p is the light-transmission of a single particle for a given wavelength, and r is the radius of the particle. It is important to emphasize that the sieve model is a good approximation only for inhomogeneous samples with non-uniform spaced particles.

The influence of the spatial pattern of particles on a measured absorbance is known as "the distributional error". The absorption statistics of the disrtibutional error effect in intensive scattering inhomogeneous objects (turbid media) was developed by Fukshansky [101]. He expressed the transmission of an ensemble of small particles as:

$$T_{sus} = \exp[-(a/b)(1-T_p)] \tag{2.9}$$

where a is thickness of a sample, and b is the mean distance between particles. Fukshansky's "distributional error" method [103] accounts for distortions of an absorption spectrum due to both multiple scattering and non-uniform spatial pattern of particles. Multiple scattering increases an apparent absorption much stronger at wavelengths with lower absorbance. The uniformly spaced particles show maximal absorption. At maximum of spectrum of non-uniform spaced particles, the absorbance is decreased.

The sieve and distributional error models operate with absorbance instead of extinction factors. They are fit for optically dense particles with sizes exceeding the wavelength of the incident light, but is not suitable for aggregates of molecules and macromolecules.

Chromophore molecules or polymers are uniformly distributed in solution over the volume; therefore, the universal Lambert-Beer absorption law (2.4) is valid. According to (2.5), the absorbance depends on the number of chromophores in the volume and does not depend on the arrangement of chromophores along the illuminated area. Therefore, the sieve effect will be negligible in solutions of molecular aggregates or macromolecules.

On the other hand, the model of the "sieve effect" can be applied to the cases where hypochromism may be an heometric-optical effect resulting from inhomogeneous light absorption in inhomogeneous media.

One would expect no optical anisotropy in a solution of macromolecules. However, this is not true for chromophores arranged in stacks along the chain (see also Chap. 3) since:

(a) the oscillator strength or the experimental extinction coefficient depends strongly on the chromophore orientation relative to the path of the incident light [61]. The maximum absorption is observed when the direction of the transition dipole moment coincides with oscillations of the electric component of the light wave;

(b) a strong anisotropy of light absorption by the films orienting the built-in dye molecules has been reported [6, 104];

(c) the absorbance of solutions of macromolecules can be changed substantially by a forced orientation [66, 105–106]; specifically, DNA is oriented by an electric field [66] and in streaming solutions [62, 107].

2.8 Light-Scattering Model

The measurement of absorbance in light scattering systems must be corrected for the effect of scattering. However, serious difficulties are encountered in attempts to explain the hypochromic spectra of polymers, specifically the spectrum of polystyrene [108], in terms of Mie's theory of light scattering encountered. This theory does not make strict physical sense: it uses the imaginary part of the complex refractive index, and the amplitude of the electric field of the light wave is identified with the intensity of the light flux [83], which is not correct. The model is a rough approximation because only the first terms of the expansion in series are accounted for. A more advanced version of this theory [108], based on the Mie model and the Beer-Lambert law, results in inadequate prediction of the dependence of the extinction coefficient on the diameter of a polymer particle.

2.9 Light-Dispersion Model

The theory of dispersion of the effective field of the light wave accounts for the local light field and the refractive index of the environment [24, 84]. Differences in intensity and in shape of the absorption spectrum of a dye in different solvents can be caused not only by molecular interactions, but also by the effects of the light field. On the other hand, the use of such poorly defined parameters as "the radius of a molecule" and "the complex refractive index of the absorption band" has resulted in a wide variety of parameters for better adjustment of experimental values. A new dimensionless parameter, called the gravity of the light field, was introduced. It is obtained by dividing the absorption parameter by the radius of the molecule and by the number of molecules per unit volume [82]. This means that the parameter is the local absorbance.

2.10 Stacking Model

The most popular model explaining the hypochromism in stacks of chromophores assumes that the oscillator strength of the band is decreased as the result of "light-induced chromophore interactions" [58, 72, 74]. Tinoco's model of light-induced chromophore interactions states that the change in the absorption during the transition of the polymer from the native to the denatured state is due to the change in the excitonic interactions between the dipoles induced in chromophores by the incident light [74]. Tinoco derived the following equation:

$$F_{0a}/f_{0a} = 1 - \frac{K\lambda_{0a}^2 f_{0a}}{N} \sum_{i=1}^{N}\sum_{j\neq 1}^{N} \left(\left[e_i e_j - \frac{3(e_i r_{ij})(e_j r_{ij})}{r_{ij}^2} \right] \frac{1}{r_{ij}^3} \right) e_i e_j - \frac{4K\lambda_{0a}^2}{N} \\ \frac{\sum_{a'\neq a}\sum_{i=1}^{N}\sum_{j\neq 1} \left(\left[e_i e_j - \frac{3(e_i r_{ij})(e_j r_{ij})}{r_{ij}^2} \right] \frac{1}{r_{ij}^3} \right) e_i e_j f_{0a}' \lambda_{0a}'^2}{\lambda_{0a}^2 - \lambda_{0a}'^2} \qquad (2.10)$$

where F_{0a} is the oscillator strength per chromophore in the polymer, f_{0a} is the oscillator strength for the monomer chromophore, $K=3e^2/8\pi^2 mc^2=1.07$, e and m are the charge and mass of the electron, respectively; c is the velocity of light, λ_{0a} is the wavelength of the absorption maximum for the monomer, N is the number of chromophores in the polymer; e_i and are e_j unit vectors in the direction of the respective transition moments in chromophores i and j; r_{ij} is the distance between the centres of groups i and j.

Tinoco's equation, derived by a formal quantum mechanical approach, can hardly be applied to real hypochromic systems. First, a strict quantitative estimation of F_{0a} for polymers and biopolymers is impossible. Second, the dipole-dipole approximation can not be applied to biopolymers, since interchromophore distances are smaller than chromophores themselves: in DNA, the purine and pyrimidine chromophores have the size of 5–7 Å, while the distances between them are ~3 Å [54]. Finally, and probably most essentially, only small (~100 cm^{-1}) excitonic effects in DNA and other biopolymer systems take place. The chromophores in proteins and DNA show almost independent behaviour: the energies of the interactions are only ~10–20 kJ/mol [109] and no intensive excitonic bands are observed. The phenomena considered in Tinoco's theory, i.e. dipole-dipole interactions between excited chromophores and migration of electronic excitation energy, in reality probably have nothing to do with the phenomenon of hypochromism: if the energy of the photon passing the polymer does not change, then there is no photon absorption and, thus, no excitation in chromophores. If the photon is absorbed by one chromophore, the neighbouring chromophores still remain in the ground electronic state.

According to Tinoco's theory, the intensity of the other bands should grow or new bands should occur. This is not observed as long as the energy of molecular interactions remains small. The model operates with oscillator strengths and does not predict the shape of the hypochromic spectrum ($h=1-F_{0a}f_{0a}^{-1}$). Besides, it does not provide a quantitative transition from theoretical quantum-mechanical parameters to the experimental ones. Tinoco's theory, developed originally for DNA [74], was repeatedly subjected to criticism [60, 75, 76], and various updates and new models [75–81, 110] were introduced, of which Rhodes' dispersion model [77], the polarization model of Bolton and Weiss [75], and the exciton model by Davydov for molecular crystals [110] deserve particular attention [60, 72, 111].

For oligonucleotides it was expected that the maximum hypochromism would be observed for di- and tri-nucleotides and would not change with the further elongation of the chain, since the strongest interactions are predicted to occur between neighbouring but not distant nucleotides. However, the maximum hypochromism can be detected for chains with 6–10 nucleotides. Moreover, the hypochromic values for di- and tri-nucleotides of different composition vary from –6% to +10%. Theory predicts that guanine-cytosine oligomers should have a larger hypochromism than adenine-thymine oligomers. Experimental data do not supports this. The strength of the guanine oscillator is greater than that of adenine. However, it appears that hypochromism of polyguanylic acid is about 0%, while that of polyadenylic acid is about 30%.

According to Tinoco's theory, the decrease in the extinction coefficients of chromophores arranged in stacks should be uniform within the entire absorption spectrum. Actually, the hypochromism within one absorption band shows its maximum at the peak of absorption, and the minimum at the tail of the spectrum. No theoretical predictions for the extinction coefficient of a chromophore in a polymer, based on the monomer extinction at a given wavelength, have been made so far (see Chapter 3).

2.11 Conclusion

The probability of absorption of a photon by a molecule at a given wavelength can be represented as a ratio of two areas, namely the absorption cross-section and the projection of electron density responsible for the optical transition. Absorption cross-section can be converted to the molar extinction coefficient. The probabilities of absorption for a number of biological chromophores were calculated.

The problem of predicting the hypochromic absorption spectra of macromolecules or molecular aggregates has not been quantitatively solved yet in the framework of traditional approaches. Most of these models, except for the one described by Bakhshiev, operate with oscillator forces and do not permit description of the form of the hypochromic spectrum. Rather, qualitative equations with adjustable parameters have to be applied to match theory with experiment.

Hypochromism of suspensions and other turbid media is interpreted from the traditional point of view of "sifting" of light fluxes between particles (the "sieve effect") or multiple light scattering by particles with special distributional absorbance statistics. Common models operate with absorbance instead of an extinction coefficient. They are suitable for suspensions of optically dense particles of size greater than the wavelength. They are hardly suitable for molecular aggregates or macromolecules.

3 Screening Hypochromism

A new model of hypochromism was suggested [68, 69] which is based on the assumption that in a molecular stack chromophores compete for the photon. This is displayed as a mutual "shielding" of chromophores from light and the effect therefore is termed "screening hypochromism". The hypochromic cross section for identical dipoles was calculated from probabilities of absorption in a stack. The screening leads to a reduction in the average extinction coefficient and to a broadening of the spectrum. Screening was considered to be insignificant initially [58] on the basis that the absorbance in a macromolecule or in a molecular aggregate is too small. However, the Lambert-Beer absorption law is hardly correct for a single macromolecule or molecular aggregate when them size smaller than wavelength of light. This law is suitable when the object is thought as a lot of layers with thickness more than the wavelength. Moreover, this low is correct only when "the molecules distributed in a finite layer are so rare that they do not shield each other" [6]. In aggregates of chromophores and in macromolecules, where distances are much smaller than the wavelength, this condition is violated. In an aggregate or macromolecule, the probability of absorption of a photon by each subsequent chromophore depends on the probability of absorption by the previous one, since they compete for the photon [63, 68]. Moreover, the screening hypochromism could be additional increased due to strong molecular interactions at very short distances.

The screening model can explain all hypochromic events that take place in solutions of polymers and biopolymers: (a) the hypochromism in chromophores arranged in stacks, (b) "residual" hypochromism in a "random coil", (c) the increase of hypochromism with the increase in the number of chromophores in a chain; (d) the dependence of hypochromism on the wavelength (variation of h within the band, the broadening of the hypochromic spectrum, maximum h at maximum absorption), (e) the lack of new absorption bands, etc. Probably, the model can also be applied to circular dichroism spectra, since the difference between the extinction coefficients for right- and left-rotatory light in macromolecules or aggregates result in a difference in the hypochromic effects for the right-handed and left-handed components. The nonconservative part of circular dichroism depends on hypochromism [112].

It should be noted that screening hypochromism is an absorption-statistical phenomenon and is not related to molecular interactions. The interactions are needed only to fix a "stack-like" arrangement of chromophores, to fix their dipole moments in parallel. The applicability of the screening model was demonstrated in case of oligonucleotides [60, 68, 69]. The model justifies the absorption-statistical nature of hypochromism (where intermolecular distances are smaller than the wavelength, but when molecular interactions can be neglected) and adjoins the models described in [82, 84, 93, 95]. A detailed consideration of the improved screening model is carried out below and its application for a number of biologically important chromophores in molecular aggregates and macromolecules is given.

3.1 Screening Model

In dilute solutions of identical molecules the probability of absorption of a quantum for each chromophore is the same: $P=P'=P''...$ (although the intensity of a light flux in solution decreases in accordance with the exponential law). In aggregates and macromolecules the probability of absorption of a quantum by each chromophore must depend on the probabilities of absorption by the other ones. When interchromophore distance is comparable with chromophore size (much less than the wavelength), one light wave covers many chromophores. This creates the conditions for competition for the given photon.

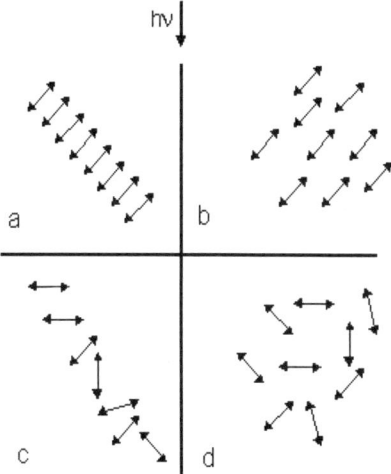

Fig. 3.1. Variants of organization of dipoles in chromophores arranged in macromolecules and aggregates: a) stack, b) shift stack, c) chaotic chain, d) chaotic aggregate

Let us consider a chain of chromophores separated by a distance much smaller than the wavelength of the incident light (Fig. 3.1). It is assumed that the vector of the electronic transition moment lies within the plane of the chromophore and that all the chromophores in the chain are identical. The stacked chromophores cause effective absorption of light hitting on the chromophore plane at right angles. In the case where the light beam is directed along the plane of the chromophore, absorption is small or absent.

As is well known from chance theory, the final probability for two accidental events is equal to $P+P'-PP'$. If the chromophores are identical, the probability of absorption (at a given wavelength) by the second chromophore will be:

$$P'=(1-P)P \qquad (3.1)$$

where P is the probability of absorption by the first chromophore. For the third chromophore:

$$P''(1-P-P')P=(1-P)^2 P \qquad (3.2)$$

For the k-th chromophore:

$$P^*=(1-P)^{k-1}P \qquad (3.3)$$

where k is the number of chromophores in the given molecular aggregate or macromolecule, provided that the distances between the chromophores are much smaller than the wavelength. The probability of absorption by the macromolecule or the aggregate of identical chromophores can be expresses as:

$$P^\wedge=P+P'+P''...+P^*=1-(1-P)^k \qquad (3.4)$$

The average probability of absorption of a photon by a chromophore within the aggregate or the macromolecule is $P^\wedge / k$. On the other hand this probability equals $2.3\ E^\wedge q / S$, as follows from (2.2). Therefore, for similarly oriented and identical chromophores,

$$P^\wedge/k=2.3E^\wedge q/S=[1-(1-P)^k]/k \qquad (3.5)$$

Thus, for similarly oriented identical chromophores we have:

$$E^\wedge=[1-(1-2.3Eq/S)^k]S/2.3qk \qquad (3.6)$$

This equation allows us to predict the hypochromic extinction coefficient in the macromolecule or the aggregate of identical chromophores based on the values of E, S, q and k. Since E changes with the wavelength, (3.6) also describes the form of the hypochromic band. This equation is valid only for regular structures, where the planes of chromophores coincide. For identical parallel dipoles at a given wavelength: $q=q'=q''=...q^*$. If the directions are different, there is no competition, and no screening is observed. Table 3.1 shows the data illustrating the change in values of $E^\wedge$ and h for different k, E and S. This Table can be used for calculating the values of $E^\wedge$ and h for stacks of regularly arranged chromophores.

Though the main contribution to the light absorption is made by the chromophores arranged perpendicularly to the light beam (see below), one should take into account the contribution of arbitrarily oriented chromophores. Thus,

$$E^\wedge=[1-(1-2.3Eqj/S)^k]S/2.3qk \qquad (3.7)$$

where j is the coefficient of contribution by the chromophores of given q to the total light absorption by systems of arbitrarily oriented stacks. In solutions of stacks, j is equal to 1.

The orientation of chromophores with respect to each other is important in the theory. With respect to the light flux the orientation is chaotic, since the molecular aggregates and macromolecules are distributed and oriented in solution chaotically as well. For the chromophores arranged in stacks, the strongest hypochromism will be observed if the directions of the dipole transitions coincide. These directions change with the wavelength [58, 62]. The largest contribution to the absorption is provided by stacks and macromolecules with the chromophore planes perpendicular to the light flux, since the dipole transitions are usually located in the chromophore plane. In a solution of aggregates or macromolecules, as well as in a solution of single chromophores, approximately 80% of light is absorbed by half of the chromophores having the value of q of about 2, and 20% by the other half with q of about 0.5. Preferentially only the first half of the chromophores participates in screening hypochromism. Therefore, for a solution of rigid regular polymers or stack aggregates with parallel transition dipoles we can assume that $q \approx 2$.

In chaotic chains (random polymers) and chaotic molecular aggregates (Fig. 3.1 c, d) only a small fraction of the chromophores is oriented perpendicularly to the light beam; it gives the maximum contribution to the absorption and participates in shielding. The majority of the chromophores compete for a photon much less efficiently due to the difference in the directions of dipole transitions. In this case, the number of chromophores competing for the photon (k) is less than the total number of chromophores in the aggregate or macromolecule (K). Chaotic chains and molecular aggregates have "residual" hypochromism. For approximate estimation of h in large aggregates and long chaotic chains ($k>100$), it can be assumed that $k=K$ with $q=1$. For short random chains and small aggregates $k<K$ always. However, we can use $k \sim K$ and $q=1$ like a first approximation. When k is too small, q may vary from 1 to 0. For this case, q should be found using exact coordinates of each chromophore in space.

Table 3.1. Hypochromic molecular extinction coefficient ($E^\wedge$) and hypochromism (h) for various values of the molecular extinction coefficient (E) and of the number of chromophores in a stack (k) at $S=10$ Å^2 and $q=1$

k	$E=0.1$		$E=1.0$		$E=3.0$	
	$E^\wedge$	h	$E^\wedge$	h	$E^\wedge$	h
1	0.1000	0	1.000	0	3.00	0
2	0.0989	1.2	0.885	11.5	1.97	34.5
3	0.0977	2.3	0.788	21.2	1.41	53.1
4	0.0966	3.4	0.705	29.5	1.08	64.1
5	0.0955	4.5	0.634	36.6	0.87	71.1
10	0.0903	9.7	0.403	59.7	0.44	85.5
20	0.0809	19.1	0.216	78.4	0.22	92.8
50	0.0598	40.2	0.087	91.3	0.09	97.1
100	0.0392	60.8	0.044	95.7	0.04	98.6

E and $E^\wedge$ are given in Å^2, and h is in percent.

For molecular aggregates or macromolecules with different types of chromophores the following general approximation may be used:

$$E^\wedge=(p+p'+p''...+p^*)S^\wedge/2.3qk \tag{3.8}$$

where $S^\wedge$ is the average effective area of the electron density, which can be approximately estimated as the arithmetic mean of all electron density areas for different types of chromophores: $S^\wedge=(S+S'+S''+...S^*)/k$; and $p, p', p''...p^*$ are the probabilities of absorption for different types of chromophores in the aggregate or macromolecule. These probabilities can be expressed as follows:

$$p=P \tag{3.9}$$

$$p'=(1-p)P' \tag{3.10}$$

$$p''=(1-p-p')P'' \tag{3.11}$$

$$p^*=(1-p-p'-p''...)P^* \tag{3.12}$$

Using (3.8–3.12) we can describe the hypochromism of a hetero-chromophore macromolecule or an aggregate of chromophores of different types. For example, for three different types of chromophores:

$$E^\wedge=(P+P'+P''-PP'-PP''-P'P''+PP'P'')S^\wedge/2.3qk \tag{3.13}$$

Equation (3.13) clearly demonstrates that although the probability of the absorption of a photon varies for different types of chromophores in the given aggregate or macromolecule, the total absorption does not depend on the position of each chromophore.

If one chromophore is shielded by a different second chromophore ("heteroscreening"), the hypochromic contribution to the absorption can be calculated using the expression:

$$E'^\wedge=E'-2.3EE'q/S \tag{3.14}$$

where $E'^\wedge$ is the hypochromic molecular extinction coefficient of the second chromophore (due to the hetero-screening by the first chromophore), E' is the initial molecular extinction coefficient of the second chromophore, E is the molecular extinction coefficient of the first chromophore, q is the orientation factor for the first chromophore, and S is the area of optical electrons of the first chromophore. The value of hypochromism, h' (divided by 100%) for the chromophore of the second type equals the value of initial probability of absorbing a photon by the chromophore of the first type:

$$h'/100\%=(E'-E'^\wedge)/E'=2.3Eq/S \tag{3.15}$$

The screening model is suitable only for small aggregates or macromolecules (less than the wavelength). This condition is important because only in this case does the probability of absorbing a photon by one chromophore depend on the probabilities of absorption by the others. If the size of an aggregate of chromophores is considerably larger than the wavelength, then the length of the elementary absorbing site approximately equals the value of the wavelength, and k is independent of the size of the aggregate. A similar situation pertains in very long macromolecules. Therefore, the value of k is defined by the number of sites in the length that does not exceed the wavelength. The screening effect can take place not only in the UV and visible regions, but also in the IR and X-ray regions.

The main features of the screening hypochromism are following: (a) variations in the hypochromism along the absorption spectrum depending on the initial extinction coefficient; (b) the most hypochromic value in the maximum of the absorption band; (c) broadening of the spectrum; (d) strong dependence of the hypochromism on the orientations of chromophores; (e) the absence of new bands or of intensities "swapping" into other bands.

In the systems where hypochromism is caused basically by molecular interactions, it is possible formally to use the obtained final equations, implying that the interactions lead to a reduction in the probability of absorption; however, the physical sense of the initial equations is changed.

3.2 Nucleotides, Oligonucleotides and DNA

For adenosine monophosphate (AMP), adenosine or adenine in aqueous solution, ε at 259 nm (maximum of the absorption band) ranges between 14 200 and 16 800 $M^{-1}cm^{-1}$ [113–115], with $\varepsilon=15000$ $M^{-1}cm^{-1}$ as the average, this value corresponds to $E=0.25$ Å^2. The projection of the valence electron density of the adenine chromophore is about 7.1Å^2 (not less than 0.9 Å^2 per projection of each of the four double bonds and 0.7 Å^2 per each of the five electron pairs on nitrogens). According to (2.2), the probability of absorbing a photon by the adenine chromophore in solution (at chaotic orientation, $q=1$) equals 0.081. In agreement with the above equations, such a P is sufficient to cause shielding in dimers.

It was demonstrated in [115] that in aqueous solution an increase in the concentration of adenine, adenosine or AMP from several µM to several mM is accompanied by a gradual reduction in ε because of the appearance of dimers, while a further increase in concentration from ~4 to ~40 mM results in a sharp fall in ε caused by the formation of large aggregates. These changes are not related to the increase in the light-scattering.

By transforming the initial and the hypochromic molar extinction coefficient of adenosine from [115] in the molecular extinction coefficient ($\varepsilon=15800$ $M^{-1}cm^{-1}=$ 0.263 Å^2; $\varepsilon^\wedge$ of dimers $=14500–15000$ $M^{-1}cm^{-1}=0.241–0.25$ Å^2, $\varepsilon^\wedge$ of aggregates $=$ about 12000 $M^{-1}cm^{-1}=0.199$ Å^2) and by substituting it in (3.6) together with $S=7.1$ Å^2 and $q=1$ one can obtain k. For small concentrations of about 0.05–0.1 mM, k equals 2 or 3, and for a large concentrations it is about 8. Thus, the equation correctly predicts that dimerization of adenine derivatives and subsequent association will take place.

Oligonucleotides are among the most thoroughly investigated hypochromic structures [113, 114]. It was determined that the arrangement of oligoadenine chromophores in aqueous solutions at temperatures below 10°C is stack-like.

Figure 3.2 depicts the absorption spectra of adenosine and penta-adenosine phosphate (pApApApApA) at the equimolar concentration of the adenine chromophore, and calculated spectrum of pApApApApA, derived from the adenosine spectrum using (3.6) under the assumption that $k=5$ and $q=2$. The k value was taken=5, because pApApApApA consists of five adenine chromophores. The $q=2$ is the average orientation factor for stacks in solution. The ε values of adenosine at each wavelength (with the 5-nm step) were used for calculation. The experimental and calculated spectra of pApApApApA are similar. The shift of the experimental spectrum in the short-wave region is 2 nm or 340 cm^{-1}, which on the energy scale corresponds to only 4 kJ/mol. A small distortion in the experimental hypochromic spectrum can be related not only to the presence of some interactions, but also to the fact that the orientation of dipole transitions in the plane of the adenine chromophore strongly depends on the wavelength [58, 62], and to the presence of a weak long-wave n,π*-transition perpendicular to the plane [109] which results in an inhomogeneity in q.

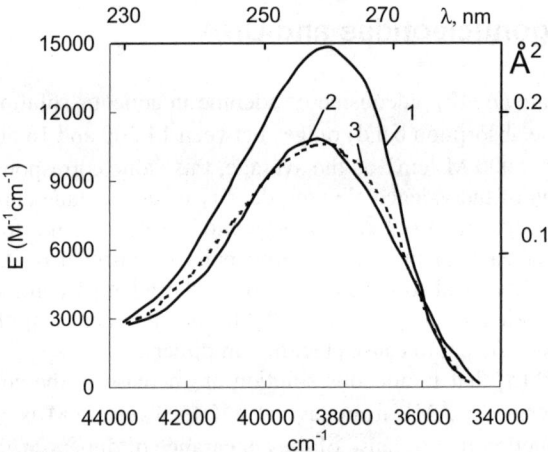

Fig. 3.2. Absorption spectra of adenosine (1), penta-adenosine phosphate (2) and the calculated spectrum of penta-adenosine phosphate at $k=5$ and $q=2$ (3). The aqueous solution (of pH 7.0) contains 100 mM of lithium chloride, 10 mM of sodium cacodylate; temperature =6°C [88, 136]

The area between the adenosine spectrum and the pApApApApA spectrum is approximately 30 times bigger than the area of the "red" hyperchromic band at 282 nm (Fig. 3.2). This means that the interactions are weak, and there is no considerable "swapping" of intensity from one band into another. The position of the "red" band is not dependent on oligoadenine length; the maximum lies at 282 nm and does not vary with k. This means that the "red" band is not excitonic or arising from super-conjugation of many chromophores. The "red" band is the result of increasing and shifting of the $n-\pi^*$ transition of adenine in polyadenines [112], probably due to intra-chain dimer formation [63]. The energy difference between the 260- and 282-nm bands is equal to about 28 kJ/mol and corresponds to the energy of adenine-adenine dimer formation. The intensity of this band depends on the number of adenine dimer pairs in the chain. As rule, no 282-nm band is observed in the absorption spectra of native nucleic acids due to the low probability of adenine-adenine pair formation in a rigid structure containing four different nucleotides.

The molar extinction coefficient of AMP at 210-nm (in the maximum of the short-wave band) is similar with that at 260 nm. The dependence of oligoadenine extinction at 210-nm on k does not differ from that at 260 nm (not shown). For example, the hypochromism at 210 nm for poly-A amounts to 28%. This agrees with the screening model, but not with models of "band intensity swapping". Similar hypochromism of the short-wave absorption band was observed for poly-A [112] and DNA [116]

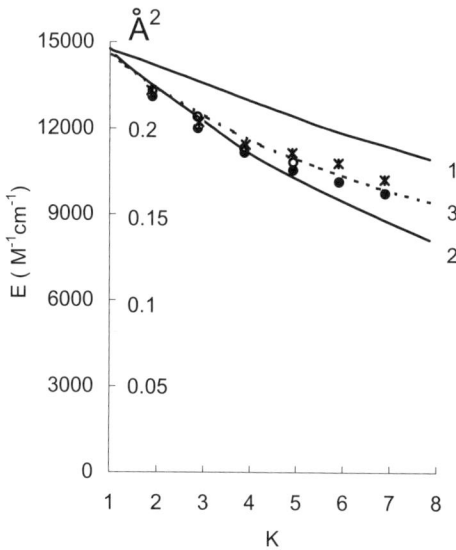

Fig. 3.3. Calculated dependencies of the extinction coefficient from the number of adenine chromophores in the chain at $q=1$ (1), in the stack at $q=2$ (2), and at $q(k) = 2 \times 0.95^{k-1}$ (3). Experimental dependences for oligoadenines from [113] at 10 °C (•), from [114] (*) and our own data (o) are presented also

Figure 3.3 depicts the calculated dependences of the extinction coefficient of adenine on the number of chromophores in the chaotic chain ($q=1$), and in the stack ($q=2$) in solution [63, 68], along with experimental data from other sources [113, 114] and our work. For small k, the experimental curve lies closely to the calculated curve with $q=2$. If $k \geq 6$, a divergence is found which is caused by the known tendency of long oligonucleotides to break and fold, i.e. by breaking of the stack arrangement. Therefore, as the oligoadenine chain gets longer, the experimental dependence approaches the calculated curve for $q=1$. The increase in temperature results in the reduction of oligoadenine hypochromism [113] because the stack arrangement is broken become virtually chaotic. At $k>10$, the calculated curve with $q=1$ crosses the experimental. It therefore can be concluded that the oligoadenine chain with $k>10$ is a chaotic chain ($q=1$). It can be easily calculated that when k is increased from 1 to 10–15, the effective q value is diminished with each k step by about 5 % from the previous value. Therefore, each $q(k)$ value may be calculated as $q(k)=q \times 0.95^{k-1}=2 \times 0.95^{k-1}$. The calculated curve with such $q(k)$ is shown in Fig. 3.3.

Let us evaluate the hypochromism of DNA. Assuming that adenine, guanine, thymine and cytosine are contained in equal proportions, the average ε at 260 nm is 8350 $M^{-1}cm^{-1}$, corresponding to $E=0.14$ $Å^2$. S of adenine $=7.1$ $Å^2$, of guanine $= 7.8$ $Å^2$, of thymine and cytosine $=5.5$ $Å^2$. An average S equal to 6.5 $Å^2$. The predominant length of the rectilinear site of DNA in solutions of high ionic strength is ≤500 $Å^2$ [109], and the length of a turn of the spiral is 34.6 $Å^2$. This means that in the site length there are approximately 14 nucleotides having orientations strictly parallel to both each other and to the electric vector of light wave. The majority of the nucleotides are rotated through different angles, which are multiples of 36 degrees, and do not participate in the competition for a photon (for a given wavelength there is a selected direction of the transition moment responsible for absorption; other directions do not contribute to absorption at this wavelength [58, 62, 109]). Only the 14 nucleotides compete for the photon, which makes $k=14$.

Substituting the known E, S and k in (3.6) where $q=2$ (a stack chain with parallel dipoles) we obtain $E^\wedge = 0.0775$ $Å^2$. From (2.6) h can be calculated to be 44.6%, which is in good agreement with the 40% hypochromism found for DNA solutions [58, 72, 109]. Stronger hypochromism may be expected in crystalline DNA because it has rectilinear domains whose lengths exceed the wavelength of light. The effective site-length where nucleotides compete for a photon is ≤260 nm. Only 75 nucleotides with identical orientations can be placed in this length; other nucleotides are rotated, so they do not compete. By substituting $k=75$ in (3.6), $E^\wedge$ is found to be 0.0188 $Å^2$. Equation (2.6) gives $h=86.6\%$. This value of h indicates the hypochromism limits of crystalline DNA.

One of a convenient dye using in studies for DNA is ethidium bromide (EB) intercalating between bases. EB bounded to DNA from phage "Lambda" has absorption bands at 300 and 335 nm (the EB spectrum is similar to that in ethanol) [117]. The 300-nm band of EB in DNA is as if "cut off" due to hypochromism [117] (see the excitation spectrum in Chapter 12). As known, the energy of "strong" interaction of EB with DNA bases is ~28 kJ/mol [118]. This energy is too small to induce the EB hypochromism. At the concentration ratio [EB]/[bases]=0.12, the 25 % hypochromism is observed [117]. EB is distributed rather uniformly along DNA [118]. At the ratio of 0.12, the dye molecules are separated by six nucleotide pairs. Therefore, any strong interactions between the EB molecules hardly possible. If calculate the hypochromism according to screening model with $E=0.23$ Å^2 (that corresponds to $\varepsilon=13000$ M^{-1}cm^{-1}), $S=10$ Å^2, $q=1$, and $k=12$ (the number of dye molecules located within 2600 Å^2), we obtain that h is equals to 25 % that coincides with the experimental value.

3.3 Tyrosine and Tryptophan in Aggregates

For tyrosine (Tyr) in water, ε is 1270 M^{-1}cm^{-1} at 275 nm [67, 119], which corresponds to 0.021 Å^2. The S of Tyr is about 3.1 Å^2 (~0.9 Å^2 per projection of each of the three double bonds and ~0.4 Å^2 per electronic pair of oxygen). Hence, $P=0.016$ (at $q=1$). This value is too small to cause shielding in dimers. However, in large aggregates shielding should take place.

Figure 3.4 shows the dependence of the extinction coefficient at 275 nm on the concentration of Tyr in aqueous solution (pH 7.0 and zero ionic strength). Dimerization and weak association lead to insignificant changes in absorption. If the concentration exceeds 2 mM, aggregates are formed, and this is accompanied by a sharp decrease in the absorption. The presence of aggregates is observed visually as an increase in light-scattering.

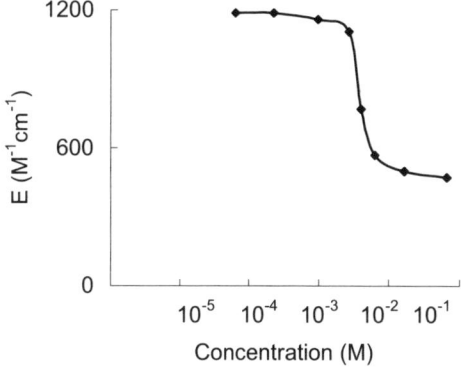

Fig. 3.4. The variation of the extinction coefficient with the concentration of tyrosine in distilled water (pH 7.0; ionic strength is ~0) [63]

Figure 3.5 depicts the absorption spectra of Tyr for high and low concentrations and the calculated spectrum, obtained for $q=1$ and $k=135$ from the spectrum of the diluted solution. The value of $k=135$ was selected for the following reasons. In large aggregates the length of an elementary shielding site is determined by the wavelength. Assuming that the average inter-chromophore distance in Tyr aggregates is 20 Å (due to the size of the amino acid), a chain of no less than 135 Tyr molecules can be fit into length of about 2700 Å. Clearly, here we discuss the order of magnitude, not the exact value of k. For the large values of k hypochromism changes not as sharply as for the average and small k. Therefore, a several-fold increase in k results in small changes in the spectrum. For example, for $k=135$ and $k=270$ the respective values of h will differ by a factor of only 1.3.

Figure 3.5 demonstrates that the calculated spectrum is close to the experimental spectrum of the concentrated solution, where Tyr is strongly aggregated. Both the 275 and 222 nm bands of Tyr are hypochromic, although to different degrees (the 222-nm band not shown). Neither new bands nor "swapping" of intensity are observed.

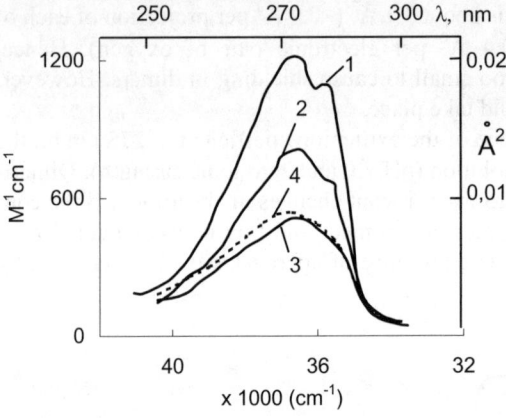

Fig. 3.5. Absorption spectra of tyrosine in water; concentrations are of 2 mM (1), 4 mM (2), 40 mM (3). The calculated spectrum (4) was obtained from spectrum (1) with $k=135$ and $q=1$ [63]. The cuvettes were placed in Specord near to detector (it allow to minimize light scattering by large aggregates due to collection of the scattered light; the residual scattering absorbance was substracted from the total absorbance using the scattering absorbance curve extrapolated from the 300–400 nm region)

In aqueous solution, the ε of tryptophan (Trp) is 5600 $M^{-1}cm^{-1}$ at 280 nm [67, 119] which corresponds to $E=0.093$ Å^2. The value of S for Trp is about 4.3 Å^2 (resulting from 0.9 Å^2 per projection of each of the four double bonds and 0.7 Å^2 per electron pair on nitrogen). Hence, P is 0.05 (for $q=1$). At high concentration Trp forms aggregates, and hypochromism is observed (Table 3.2).

3.4 Tyrosine and Tryptophan in Proteins

Proteins are much smaller (10–100 Å) than the wavelengths of the absorption bands of Tyr and Trp. The peak value of P in the absorption band of Tyr equals 0.016 and of Trp is 0.093. Due to the large number of aromatic amino acids in proteins, efficient shielding should be expected. With increasing number of Tyr and Trp more dipoles will be parallel to each other, and the shielding will be stronger.

3.5 Tyrosine and Tryptophan in Aggregates

Consider two proteins (of known crystal structure), in which hypochromism has been observed [119]: bovine phospholipase A2, containing one Trp and seven Tyr residues, and ribonuclease Tl, containing one Trp and nine Tyr residues.

Equation (3.6) allows us to take into account the effect of self-shielding of Tyr at 275 nm, where $E=0.021$ Å^2 (as for Tyr in aqueous solution), $q=1.5$, $S=3.1$ Å^2 and $k=7$ or 9 for phospholipase A2 or ribonuclease Tl. The choice of $q=1.5$ is due to the fact that more than half of the Tyr residues in these proteins are nearly parallel to each other (in this case the increase in q from 1 to 2 makes only a small difference). For tyrosines in phospholipase, $E^\wedge=0.0196$ Å^2 and $h=6.7\%$; for tyrosines in ribonuclease, $E^\wedge=0.0191$ Å^2, $h=8.9\%$. Thus, self-shielding of tyrosines does take place, but the effect is small.

By using (3.14) and (3.15) the hetero-screening of tyrosines by Trp can be calculated. We take E' equal to $E^\wedge$ of tyrosines (the only known value), E of Trp=0.09 Å^2 (as for Trp in solution at 275 nm), S of Trp $=4.3$ Å^2, $q=1$ for phospholipase (since Trp in phospholipase rotates freely [119]) and $q=1.5$ for ribonuclease (because of the orientations of the Trp and Tyr). Hetero-screening of the tyrosines by Trp is 11% in phospholipase and 16 % ribonuclease.

Hence, for Tyr hypochromism, after taking into account both self-screening and hetero-screening by Trp, are ~18% for phospholipase and ~25 % for ribonuclease. These values of h are two times smaller than the experimental values of h in [119], and the question remains for the reason for such a difference.

First, in [119] it was assumed that only the Tyr band is hypochromised, and the difference between the spectrum of the protein and that of Trp was attributed exclusively to the tyrosines. However, hypochromism of Trp due to being hetero-screened by tyrosine is likely to occur as well. Therefore, in the differential spectrum a measurable contribution by Trp is expected to be present. Secondly, normalization of the spectra at the red edge is prone to error, because the red-edge sites of Trp band can change strongly on going from the aqueous phase to a protein. Third, interactions of Trp or Tyr residues with any neighboring groups also change the absorption spectra. By not taking into account the above considerations, the value of hypochromism of Tyr in [119] has been overestimated.

Hetero-screening of Trp by tyrosines can be evaluated using the following procedure, since (3.14) and (3.15) can not be directly applied in this case. By neglecting individual differences between tyrosines in q, an approximate value of the total probability of the photon absorption by tyrosine residues can be found from (3.5). For phospholipase it equals 0.107, and for ribonuclease it equals 0.135. Substituting these, values for p and the value of 0.048 for P' (the P value of Trp at 275 nm) in (3.10), we obtain $p'=0.043$ for phospholipase and $p'=0.041$ for ribonuclease. Here p' is the probability of photon absorption by Trp shielded by tyrosines. By substituting the values of p' for P in (2.2) and using the values of $S=4.3$ Å^2 and $q=1$ for Trp, the extinction coefficient of Trp in phospholipase at 275 nm is found to be 0.08 Å^2, and that in ribonuclease 0.077 Å^2. Thus, the extinction coefficient of Trp at 275 nm is reduced by 0.01 Å^2 in phospholipase and by 0.013 Å^2 in ribonuclease, i.e. the h of Trp is 11.1% and 14.4% respectively.

The ε of phospholipase at 275 nm is equal to the sum of the hypochromic extinction factors of seven tyrosines and one Trp, i.e., 7×0.0187 Å^2+0.08 Å^2=0.211

Å² (the calculated $E'^{\wedge}$ for Tyr in phospholipase is 0.0187 Å²). Similarly, for ribonuclease we obtain E=0.236 Å² (the $E'^{\wedge}$ for Tyr in ribonuclease is 0.0177 Å²). Assuming that in the complete absence of hypochromism the value of E of phospholipase should be 0.237 Å², (0.09 Å²+7×0.021 Å²), and of ribonuclease is 0.279 Å², we obtain (2.6) that h of phospholipase at 275 nm is about 11% and for ribonuclease it is about 15%.

3.5 Hemoglobin in Erythrocytes

For one hemoglobin sub-unit in the Soret band ε is 120 000 M^{-1}cm^{-1} [60, 120] which corresponds to ~2 Å². The value of S for heme is approximately 20 Å². Therefore, the P value for one sub-unit is about 0.23 (for q=1).

A strong hypochromism in the Soret bands of aggregates of heme-containing octa-peptides isolated from cytochrome-c was described in [120]. On increasing the concentration by about 1000 times, a large hypochromism of aggregates occurred, with h as high as 50% at the peak of the Soret band. Based on circular dichroism experiments, the results were interpreted in terms of exciton resonance. However, no quantitative calculations for the hypochromic absorption spectra were made. By substituting the initial absorption spectrum of such a heme-octa-peptide in (3.6), spectra similar to the hypochromic spectra presented in [120] are obtained (not shown here). Probably, the hypochromism of heme-octa-peptide aggregates can be explained by shielding.

In erythrocytes the Soret band is strongly hypochromised and wide as compa-

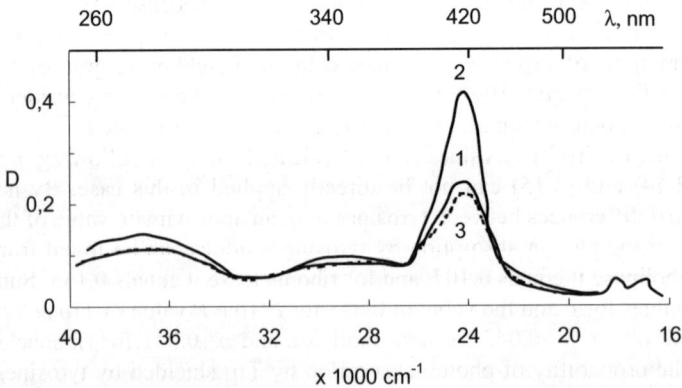

Fig. 3.6. Absorption spectra of rat erythrocyte suspension in 150 mM NaCl solution (1), hemolysate obtained by hemolysis of erythrocytes in distilled water (2), and the calculated spectrum for k=7, q=1 (3) [63]. The cuvettes were placed in Specord near to detector (the residual scattering absorbance was substracted from the total absorbance using the scattering absorbance courve extrapolated from 700 nm)

red to that of the hemolysate solution (Fig. 3.6) where the absorption basically is caused by free hemoglobin. The h values varies along the spectrum and strongly depends on the ε of hemoglobin. Maximum hypochromism takes place at a maximum of the Soret band. Hypochromism at a maximum of the Soret band varied from 40 to 70 % depending on experimental conditions (purity of an erythrocyte fraction, resistance to hemolysis, etc.). The low-absorbing bands in the range of 530–590 nm are almost not hypochromised, in conformity with the model. Band "swapping" and new bands are not observed. These characteristics confirm that the hypochromism of hemoglobin in erythrocyte suspension is a heometrical-optical effect, probably caused by the shielding of light in each single erythrocyte.

Since the size of an erythrocyte is larger than 10000 Å, the size of an elementary site, in which hemes compete for a photon, is determined by the wavelength. It equals approximately 4000 Å for the Soret band. Erythrocytes contain a 34% hemoglobin solution, which corresponds to a concentration of 5 mM (the molecular weight of the protein is 68000). For such a concentration, the center to center inter-molecular distances are about 100 Å. It allows the stacking of about 40 protein globulae in the length of 4000 Å. Therefore, k cannot exceed 40. Since it is unknown whether the hemoglobin in erythrocyte is evenly distributed or forms aggregates smaller than the wavelength, the exact value of k cannot be calculated precisely. The experiment is in the best agreement with theory if k ranges from 7 to 20, depending on the test parameters and experimental conditions. The spectrum of fresh hemolysate (Fig. 3.6) obtained from (3.6) with $k=7$, $S=20$ Å^2 and $q=1$ does not significantly differ from the experimental spectrum of erythrocytes. Hypochromism at the peak of the Soret band varies from 40 to 70% depending on the conditions of the experiment (Fig. 3.6 and Table 3.2). Low-absorbing bands in the range of 530–590 nm are practically not hypochromised, which is in good agreement with the theory.

The differences we find between calculated and experimental hypochromic spectra can be caused by the partial hemolysis of erythrocytes in suspension, by the contributions from other proteins and coenzymes, by the dependence of the directions of heme dipole transitions on the wavelength, etc. The contribution from the "environmental polarization" cannot be significant, because the refractive index of water differs little from that of the erythrocyte matrix. The scattering absorbance of the erythrocyte suspension did not exceed 0.05. Absorbance at the Soret band was about 0.4–1.5. Therefore, light scattering does not play a significant role here as well (a small contribution of the light scattering by erythrocytes is reduced by extrapolating the "scattering absorbance" from 600–700 nm).

3.6 Chloroplasts and Visual Rods

In work [102], an experimental-theoretical procedure was described which permits the derivation of the authentic chlorophyll spectrum by separating the in *vivo* absorption spectrum of a leaf from distortions resulting from multiple scattering and distributional error. This was a new interpretation of sieve effect. Authentic absorption spectra on scale of absorbance were calculated, however, spectra on the scale of extinction coefficient were not obtained. So, let us calculate the hypochromism of chlorophyll in chloroplasts using the screening model.

The ε of chlorophyll at the peak of absorption, 430 nm, is about 120000 M^{-1} cm^{-1}, which corresponds to 2 $Å^2$. S reaches 17 $Å^2$, and $P=0.271$. Equation (3.6) predicts that even in the case of a stack dimer ($k=2$, $q=2$) a notable hypochromism in the Soret band is likely; $h=27\%$ at 430 nm. This is in agreement with the experiment.

It is well known that chloroplasts are organized in a stack. The absorption spectrum of chloroplasts is strongly hypochromised as compared to the spectrum of free chlorophyll. For example, in the spectrum of a Chlorella suspension an ~52% hypochromism was observed at 440 nm [93] when compared to the methanol solution of chlorophyll extracted from it. For $h=52\%$, we obtain from (2.6) and (3.6) that the values of k are approximately 4 ($q=2$) or 7 ($q=1$). These k values are in good agreement with the known chlorophyll contents in the pigment-protein light-harvesting complexes.

The size of thylacoids (the elementary units of which chloroplast grains are built up) is about 5000 Å, which is close to the wavelength. A stack of about one 1000 chlorophyll molecules can be accommodated in such a length. This can result in more than 99% hypochromism in the Soret band and "daubing" of the spectrum. Usually this is not observed. Probably, it is related to the availability of lots of the freely dissolved chlorophyll in chloroplasts and also to the fact that some chlorophyll molecules form separate pigment-protein complexes of small sizes and different orientation.

For rhodopsin, the ε of retinol at the peak, 498 nm, is about 40000 $M^{-1}cm^{-1}$ [58] which corresponds to nearly 0.7 $Å^2$. The S value of retinol can be estimated to be 5.4 $Å^2$. That makes $P=0.3$. The high P value of which is indicative of the fact that in the structures of retinol aggregates such as visual rods and in membranes containing bacteriorhodopsin substantial shielding may take place. As much as 30% hypochromism can be reached for a stack dimer ($k=2$, $q=2$). For 10 rhodopsin molecules organized in a chaotic aggregate, h is 67%, whereas in a stack h is 83%. A significant difference between the absorption spectrum of rhodopsin and the action spectrum of rod sight was observed and explained by self-shielding in the outer segments of the visual rods [121].

3.7 Clusters of Aromatic Hydrocarbons

Screening hypochromism should be observed in molecular microcrystals. We can verify this by using anthracene as an example. An evaluation of the probability of photon absorption by anthracene was founding [67]. Here we perform more strict calculations for anthracene using the value of $S=8$ $Å^2$ (greater than 1.1 $Å^2$ per each of the seven double bonds) along with more accurate values of ε.

The highest ε of anthracene is found for ethanol or hexane solutions. At 251–253 nm they are 200 000 $M^{-1}cm^{-1}$ [122], which corresponds to 3.3 $Å^2$. Therefore,

$P=0.95$. The large value of P is indicative of the fact that almost every photon becomes absorbed. The transition in the 252-nm band can be described as a circular oscillator, i.e. q never exceeds 1. At 375 nm, $\varepsilon=7500$ M^{-1}cm^{-1} [122], which corresponds to 0.125 Å^2. This results in a P value of 0.036. We can assume that in a microcrystal of anthracene the 253-nm band should be hypochromised very strongly, but the band at 375 nm will be hypochromised much less than that, the value of the hypochromism being dependent on the thickness of the microcrystal.

In anthracene clusters, which may form in aqueous solution when small quantities of ethanolic solution of anthracene are added, a very strong hypochromism of the 253-nm band is observed, and the hypochromism observed in the long-wavelength bands is much smaller (Fig. 3.7 and Table 3.2). A "red" shift of 1200 cm^{-1} in the long-wavelength bands in water arises due to the strong molecular interactions in anthracene clusters. The form of these bands changes as well. The absorption spectra of anthracene dimers and microcrystals are presented in [123, 124]. The spectrum of anthracene in water is very similar to that of anthracene microcrystals but not to that of dimers. This again provides evidence that in aqueous solution the anthracene is clustered even at a very low concentration, while the dimers are absent.

The sizes of the anthracene clusters do not exceed 100 Å, i.e., k can not exceed

Table 3.2. Extinction coefficients and hypochromism in chromophore aggregates and macromolecular structures

Structure	k	q	$E^\wedge$ [Å^2]	$\varepsilon^\wedge$ [M^{-1}cm^{-1}]	$\varepsilon\#$ [M^{-1}cm^{-1}]	h [%]	$h\#$ [%]
PApApApApA	5	2	0.1811	10860	11100	27.6	26
DNA in solution	14	2	0.0775	4650	-	44.6	40
DNA in crystal	75	2	0.0188	-	-	86.6	-
Ethidium in DNA	12	1	-	9750	9750	25.0	25
Tyr in aggregates	135	1	0.0088	528	530	58.2	58
Trp in aggregates	30	1	0.0488	2940	2800	47.5	50
Tyr and Trp in phospholipase A2	8	-	-	-	-	~11	-
Tyr and Trp in RNAase T1	-	-	-	-	-	~15	-
Hemoglobin in erythrocytes	20	1	0.432	25920	36000	78.4	70
Chlorophyll in Chlorella	7	1	0.94	56400	57600	53.0	52
Rhodopsin in rods	10	1	0.228	13680	-	67.4	-
Anthracene in clusters	10	1	0.3478	20870	20700	89.5	90

$\varepsilon\#$ and $h\#$ are the experimental hypochromic values, $E^\wedge$ and h are calculated ones.

20–30. The calculated spectrum of anthracene microcrystals, obtained from the spectrum of anthracene in ethanol solution for $k=10$ using (3.6), is shown in Fig. 3.7. In the short-wave region it looks very similar to the spectrum of anthracene clusters. In the long-wavelength region, a qualitative similarity is observed with respect to intensity, but not for the positions of the bands. This is caused by molecular interactions.

The aqueous solutions of anthracene were used contained about 0.3% ethanol. Therefore, not all of the anthracene molecules are clustered and the absorption values are somewhat overestimated. Nevertheless, h at 253 nm equals 96 % for $k=30$, which does not differ much from the value of h for $k=10$. Thus, even in the case of strong molecular interactions and using rough estimates for k, the theory is in good agreement with the experiment.

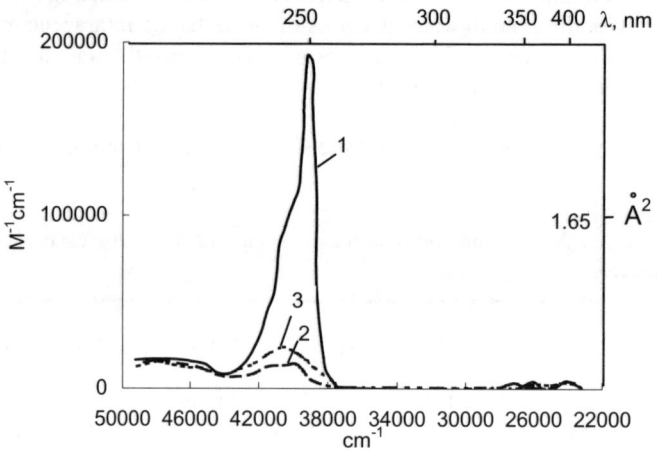

Fig. 3.7. Anthracene spectra in ethanol (1), and water (2), and the calculated spectrum for $k=10$, $q=1$ (3) [63]

Conceivably, the screening model may serve to explain the known hypochromism [125] in the 253-nm band of anthracene chromophores in poly(anthrylmethacrylate) copolymers. Figure 3.8 shows the strong hypochromism in the 250-nm band and a small hypochromism in the low-intensity bands of longer wavelengths. The increase in the absorbance at the edges of the 250-nm band is probably due to light scattering and chemical changes in the anthracene chromophore attached to a polymer. In the copolymer, the anthracene chromophores form sandwich pairs [125]. By using (3.6) for such a dimer ($k=2$), the hypochromism (h) at the peak of the 250-nm band can be calculated to be about 50%. This value coincides with the experimental data (Fig. 3.8). In the long-wavelength region only small changes take place. This is in agreement with our explanation. A small "red" shift in the spectrum of the copolymer is likely result from molecular interactions.

An attempt of theoretical description of absorption behavior for molecular dye

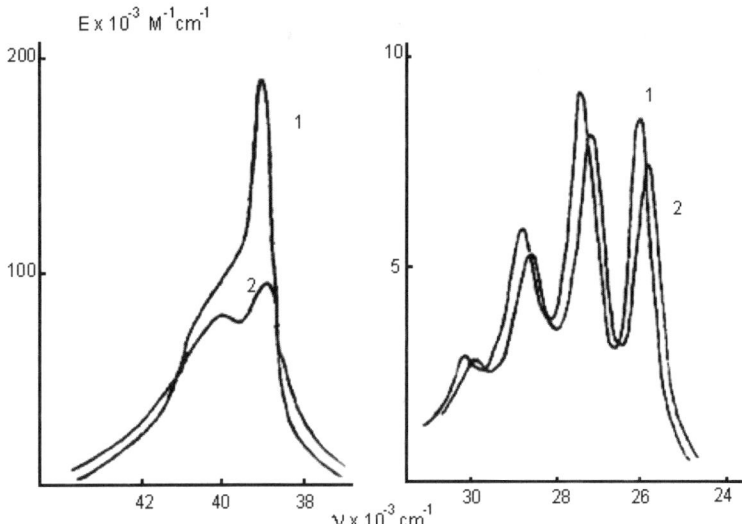

Fig. 3.8 Absorption spectra of the anthracene chromophore in 9-anthrylmethylacetate (1) and in the 9-anthrylmethylmethacrylate/methylmethacrylate (0.54/1.0) copolymer (2) in dioxane solution at 20 °C [125]

aggregates like dense spheres, loosely packed spheres, straight chains, and coiled chains was done also in [526].

3.8 Conclusion

This chapter has demonstrated that mutual screening of chromophores due to competition for the incident photon can take place in stack-like aggregates and macromolecules. Screening leads to a decrease in the extinction coefficient. The most decrease takes place in the maximum of the absorption band. Using adenine, tyrosine, tryptophan, porphyrin and other chromophores, it was shown that in oligonucleotides, nucleic acids, proteins, cells, microcrystals and polymers the mutual shielding of chromophores from light can take place, which results in a decrease in the extinction coefficient and in the broadening of the absorption band. The model allows to calculate the hypochromic absorption spectra directly from the extinction coefficients of single chromophores, without using the imaginary part of complex refractive index. The model correctly predicts hypochromic absorption spectra and allows for the evaluation of the number of chromophores in aggregates and macromolecules provided that molecular interactions are negligible. Also, the screening model is capable of determining the orientation of chromophores in macromolecules or aggregates from the absorption spectra.

4 Photometric Estimation of Protein Content in Biological Suspensions

4.1 Colorimetry and Others Methods

The quantitation of protein in suspensions of biomembranes and cells is one of the most topical problems in biological research. Colorimetric methods based on a chemical reaction between a reagent and protein are extensively applied [126, 127]. The most widely used methods are the biuret reaction and the Lowry assay [127, 128]. Unfortunately, many of the compounds used in enzyme purification interfere with the reaction. To overcome this problem, modifications are required [127]. Furthermore, the accuracy of these methods depends on the number of Trp and Tyr residues in a protein. Other methods are based on the change in the absorbance of certain dyes, which bind to proteins [129–131]. In these methods the resulting color depends, to a limited extent, on the composition of the protein, therefore, accurate protein quantification requires a calibration using an appropriate protein mixture [127].

While being highly sensitive and relatively specific, these approaches also have certain drawbacks: they are labor consuming, destroy the protein, and require unstable reagents [126, 127, 130, 131]. The spectrophotometric method of measuring concentrations of individual water-soluble proteins is free from the above limitations [132, 133]. However, it has not been applied to the studies of membranes and cells since the numbers of Trp and Tyr residues, essentially contributing to the absorption at 260–290 nm, varies significantly from one protein to another. This makes the conversion from the absorbance (optical density, D) to protein concentration impossible. Moreover, the analysis is complicated by the effects of hypochromism and light scattering in membranes and cells.

4.2 Combined UV-Spectrophotometric Method

An attempt was made [134] to find such a combination of the colorimetric and UV-spectrophotometric methods that would enable one to perform a rapid and accurate estimation of protein concentrations in biological suspensions.

Bovine serum albumin (fraction 5, "Boehringer"), sodium dodecyl sulfate

("Serva"), mitochondria and sarcoplasmic reticulum were used in our study. Mitochondria were isolated from the liver of "Wistar" rats by a conventional procedure [133]. Male rats of 250–300 g, were used. One animal was taken for each isolation. Sarcoplasmic reticulum was isolated from rabbit muscles by the method described in [135] (the isolation was carried out at the Biology Department of the Moscow State University). The "light" and "intermediate" fractions of sarcoplasmic reticulum were used (LSR and ISR, respectively); ISR is sensitive to caffeine. The original membrane suspension was thoroughly homogenized and diluted in a 1-cm quartz cuvette so that its absorbance at 280 nm did not exceed 1. Measurements in the UV range were carried out on the "Specord" spectrophotometer (Germany) in the compartment for light-scattering samples. A part of the suspension was used to measure the protein concentration by the biuret reaction or by the Lowry assay [127, 128]. Albumin solution was used as a standard in colorimetric measurements. To eliminate the effects of Tris buffer and sucrose on the Lowry reaction [131], mitochondria were suspended in distilled water. For more accurate measurements the mitochondrial suspension was treated with a detergent, sodium dodecyl sulphate (SDS). The sarcoplasmic reticulum was diluted in a medium containing 2.5 mM of magnesium chloride, 100 mM of sodium chloride, and 5 mM of imidazole, pH 7.0.

Table 4.1 represents the absorbance data at 280 nm for mitochondrial suspensions and for the two fractions of sarcoplasmic reticulum with various protein concentrations, determined by the Lowry assay for mitochondria and by the biuret reaction for reticulum. As is seen, UV-absorption makes it possible to estimate protein concentration in the range from 0.05 to 0.5 mg/ml. Using spectrophotometers with absorbance from 0.1 to 3.0 the concentration range could be extended to 0.005–1.0 mg/ml.

Table 4.1. Protein concentration and absorbance of suspensions at 280 nm [134]

Colorimetric concentration [mg/ml]	Absorbance at 280 nm:		
	Mitochondria	LSR	ISR
0.05	0.10	0.17	0.16
0.10	0.20	0.34	0.32
0.50	0.97	-	-

For the sarcoplasmic reticulum membranes, the value of absorbance at 280 nm corresponds to the light absorption by Trp and Tyr residues (mainly of Ca^{2+}-ATPase, Chapter 14), because reticulum membranes contain mostly proteins and lipids, where lipids do not absorb at 280 nm.

Mitochondria are known to contain nucleotides, ubiquinone and other substances absorbing in the UV-range. Moreover, the mitochondrial suspension strongly scatters the light. Thus, the UV-absorption method should be thoroughly considered.

Fig. 4.1 demonstrates the UV-absorption spectra of mitochondrial suspensions

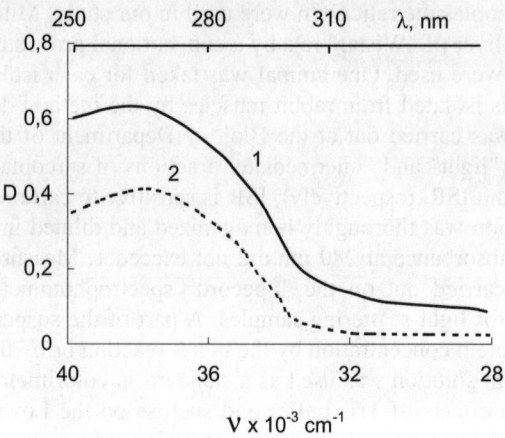

Fig. 4.1. UV absorption spectra of: mitochondrial suspension (1), solution obtained by treating mitochondria with SDS (2). The concentration of SDS is 4% [134]

and the solution obtained by treating mitochondria with a detergent, sodium dodecyl sulphate. The absorption peak is located at 262 nm, which is typical for nucleotides and ubiquinone. Treating mitochondria with SDS drastically reduces light scattering (at 320 nm, where Trp and Tyr residues do not absorb the light, and mitochondrial suspension has obvious absorbance due to scattering). It is well known that proteins make a major contribution to the UV-absorption of mitochondria at 280–290 nm [15]. Therefore, if mitochondria are isolated by the conventional method and then treated by SDS, the ratio between the value of absorption at 280 (or 290) nm and the colorimetrically defined concentration is sufficient to

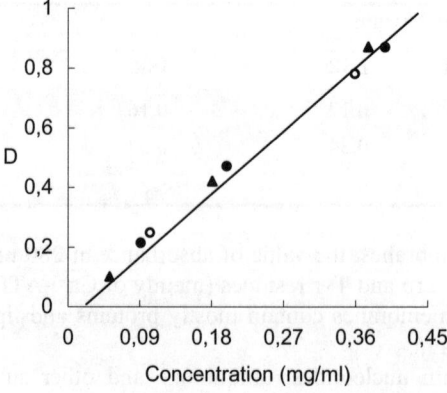

Fig. 4.2. Dependence between absorbance at 280 nm for SDS-treated mitochondria and the colorimetrically estimated protein concentration of mitochondria [134]

be once determined, so that further measurements would involve only UV-recording. Concentration would be estimated by the calibration curve. The content of non-protein components in mitochondria can vary from one rat to another, but these variations do not affect the absorbance at 280-290 nm (Fig. 4.2).

Fig. 4.2 shows the dependence of the absorbance at 280 nm of the SDS-treated mitochondrial solution on the protein concentration estimated by the Lowry procedure. The data from three sets of measurements (three suspensions of the mitochondria obtained from three different rats) are shown. As is seen from this Figure, points are not scattered but form one line. The mean experimental error do not exceed 3.5% and is mainly due to the error in the Lowry method.

4.3 Comparison of the Methods

Colorimetric methods (the Lowry assay and biuret reaction) from [126, 127, 136] and the combined UV-spectrophotometric method [137] are compared in Table 4.2.

Table 4.2. Comparison between the combined UV-spectrophotometric method and colorimetric methods

Parameter	Lowry procedure	Biuret reaction	Combined UV method
Object under study	Peptide bonds and aromatic residues	Peptide bonds	Aromatic residues
Sensitivity, [mg/ml]	0.05–0.5	0.5–5	0.005–1.0
Protein destruction	Yes	Yes	No
Interfering substances	Tris-HCl, abscorbate, sucrose, etc.	Ammonia	Substances with absorption at 280–290nm
Accuracy of method	±2–±10%	±10%	±2–±10%
Reagents	Folin reagent	Copper	Without or SDS
Storage time of reagents	1–2 months	1–2 months	Unlimited
Coloring rate	Slow	Fast	Momentary

Table 4.2. (cont.)

Stability of coloring in time	Several hours	Hours or day	Unlimited time
Time required	50–60 min	20–30 min	3–5 min

The Table demonstrates that the use of the combined method for the conventionally prepared suspension leads to the results that could be hardly obtained by the colorimetric methods. The accuracy of the colorimetric methods depends on type of membranes, temperature, presence of sucrose, lipids etc. Furthermore, the reagents used are not stable and the color is changed with time. Therefore, the colorimetrically determined protein concentrations could considerably vary from one experiment to another. For instance, protein concentrations in the same original preparations of mitochondria or reticulum obtained with the use of the colorimetric methods in seven laboratories of our Institute were differ by 10–50%. The combined method lacks these drawbacks. All the concentrations of membranes given in this book were found by the combined UV-spectrophotometric method.

4.4 Conclusion

A combined UV-spectrophotometric method of measuring the amounts of proteins in conventionally isolated biological suspensions is proposed. The method is based on the comparison between data on absorbance of a suspension or a SDS-treated solution at 280 or 290 nm and the results of a single colorimetric determination. The combined spectrophotometric method reduces the time of measurement, has higher specificity, sensitivity, and accuracy as compared with the traditional colorimetric methods. Calibration curves are obtained for conventionally isolated rat liver mitochondria and two fractions of sarcoplasmic reticulum.

5 Screening and Reabsorption of Light

5.1 Inner Filter Effect

To measure luminescence in solutions, screening and reabsorption of light should be thoroughly taken into consideration [138]. The sum of screening and reabsorption was named as "inner filter effect". Usually correction to the inner filter effect is calculated by the approximate expression [67]:

$$F_{corr} \sim F_{obs} \text{antilog}(D_{ex}/2 - D_{em}/2) \quad (5.1)$$

where F_{corr} is the corrected fluorescence intensity, F_{obs} is the observed intensity, D_{ex} and D_{em} are the optical densities (absorbances) at the excitation and emission wavelengths. The use this expression as an exact equation sometimes leads to mistakes. For instance, when it was made in [139], the excitation spectrum of tyrosine fluorescence in the protein-DNA complex in the range between 250 and 270 nm was distorted due to a high DNA absorbance. Similar mistake contains in the book [58]: the excitation spectrum of ethidium in the presence of DNA or RNA is totally cut off below 280 nm (see a correct spectrum in Chapter 12).

In weakly absorbing solutions (where the absorbance is less than 0.01) the effects of screening and reabsorption are not essential. In weakly absorbing heterogeneous systems, e.g. diluted biological suspensions, these effects are usually not taken into account. This is not always correct.

When single molecules form large aggregates or when they get incorporated into particles of a suspension, absorbance (optical density, D) for the entire system may not change or insignificantly changes (due to sieve effect). However, local absorbances (D_L) in this heterogeneous system can be essentially higher than absorbance of the entire system (Fig. 5.1). Thus, local screening and reabsorption, i.e. microscreening and microreabsorption, take place and result in a decrease in the luminescence of molecules in particles [60, 140]. These phenomena are especially revealed in particles of a large size (d), when $d \gg \lambda$.

Reabsorption and reemission processes in thick and thin layers and in lateral microstructured samples in the presence of some aromatic compounds (perylene on silica, etc.) were studied in [141]. Photons emitted by the "donor" molecules in a micron-size spherical particle remain inside the particle for a long time, that leads to an increase in the probability of reabsorption by the "acceptor" molecules in this particle [142].

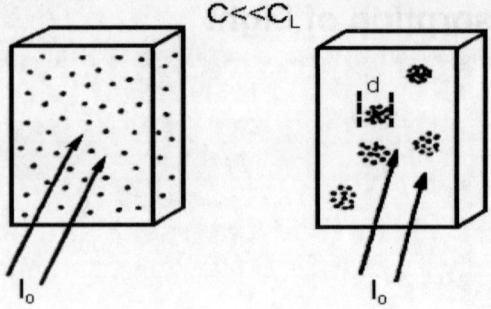

Fig. 5.1. Formation of high local absorbances in particles. Left, solution of molecules; right, suspension

5.2 Microscreening and Microreabsorption

Let consider a diluted suspension of identical particles. The intensity of luminescence can be given in the following equation (here hypochromism and scattering of light are not taken into account):

$$I_F(\lambda) = I_o(\lambda) \, Q_f[1 - T_F(\lambda)] \tag{5.2}$$

where I_o is the intensity of the exciting light, Q_f is the quantum yield of luminescent molecules in the particles, T_F is the transmittance of the suspension. This equation is similar to the one describing luminescence in a homogeneous molecular solution. If an absorbing substance, e.g. a dye, is introduced into the suspension and becomes uniformly distributed over the volume, then the intensity of luminescence from the particles will be reduced due to the screening and reabsorption of light (same as in a homogeneous molecular solution) The reduced intensity is:

$$I_S(\lambda) = I_F(\lambda) T_R(\lambda) \tag{5.3}$$

where T_R is the transmittance of the dye solution, $T_R(\lambda) = 10^{-D_R(\lambda)}$, $D_R(\lambda) = E_R(\lambda) C_R l$, where E_R is the extinction coefficient of the dye in the wavelength range of excitation or emission of luminescence, C_R is the final concentration of the added dye, l is the optical path in the suspension. For an infinitely thin layer $l\sim 0$ we have $T_R \sim 1.0$ Therefore, the effects of screening and reabsorption could be neglected. However, if molecules of the dye selectively bind to particles, the local concentration of the dye inside a particle (C_L) is much greater than the mean concentration (C_R), i.e. $C_L \gg C_R$. As a result, the local absorbance (D_L) of the dye inside each particle will be greater than the mean absorbance in the suspension, $D_L(\lambda) > D_R(\lambda)$ [60, 140]. The value of C_L is given by the equation:

$$C_L = C_R \gamma / VnN \tag{5.4}$$

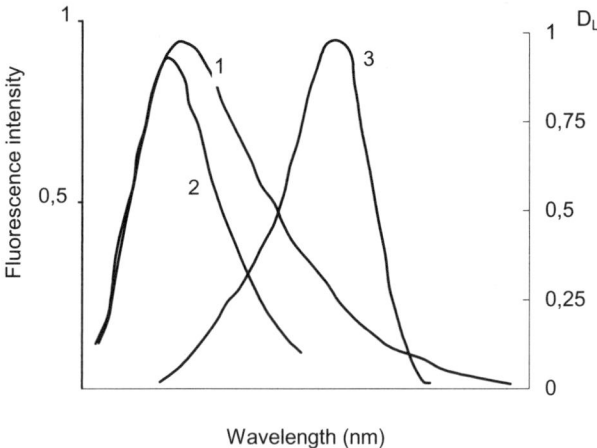

Fig. 5.2. Spectral changes resulting from microscreening and microreabsorption: (1) the original spectrum of excitation or emission; (2) the spectrum of excitation or emission distorted by microscreening or microreabsorption; (3) the absorption spectrum of the dye. Spectrum 2 was calculated from the spectra 1 and 3 with $D_L=1.0$ at the peak of the absorption band [140]

where n is the number of moles of particles in a unit of illuminated volume, i.e. the molar concentration of particles, V is the volume of one particle, N is Avogadro's number, γ is the coefficient of the dye partition between the particles and the solution.

Consider the case where all molecules of the dye penetrate into the particles. In this case the intensity of luminescence from the suspension will be:

$$I_M(\lambda)=I_F(\lambda)T_L(\lambda)T_B(\lambda) \qquad (5.5)$$

where T_L is the transmittance of a single particle, T_B is the transmittance of the entire suspension.

Assuming that the Lambert-Beer law is valid for each particle whose size is much greater than the incident wavelength, we can write [60, 140]:

$$D_L(\lambda)=\varepsilon_L(\lambda)dC_L \qquad (5.6)$$

where ε_L is the extinction coefficient of the dye in the particle, and d is the effective length of a single particle. If $\gamma=1$, then $T_B(\lambda)=10^{-\varepsilon_L(\lambda)C_R l}$. For an infinitely thin layer $T_B \sim 1.0$. In this case we have [60, 140]:

$$I_M(\lambda)/I_F(\lambda)=T_L(\lambda)=10^{-\varepsilon_L(\lambda)C_L d} \qquad (5.7)$$

Thus, when microscreening and microreabsorption take place, the intensity of luminescence and the shape of the excitation and emission spectra can be easily calculated if $\varepsilon_L(\lambda)$, C_L and d are known or $D_L(\lambda)$ can be determined experimentally for a single particle. Unlike trivial macroscopic screening and reabsorption, the effects of microscreening and microabsorption are independent of l. However, microscreening and microabsorption result in the same decrease in the intensity and distortion in the shape of a luminescence spectrum. When the absorption spectrum of the dye overlaps the spectrum of emission from the particles, microreabsorption is observed. In the case where the absorption spectrum of the dye overlaps with the absorption spectrum of the particles, microscreening is observed (Fig. 5.2).

5.3 Erythrocytes and Other Cells

Microscreening and microabsorption can be observed in smaller particles if C_L is large. Similar effects can be observed without additional staining in the particles that naturally contain substances having a large extinction coefficient in the range of luminescence excitation or emission. Let us estimate microreabsorption of protein luminescence by heme in the Soret band in a single erythrocyte. Erythrocytes are known to contain a 34% solution of hemoglobin. Taking into account that the molecular mass of hemoglobin is 68 000, the local concentration of hemoglobin is equal to $\sim 5 \times 10^{-3}$ M. The minimum thickness of an erythrocyte is 10^{-4} cm. Therefore, the optical path for microreabsorption should not be less than 5×10^{-5} cm. The extinction coefficient of hemoglobin tetramers at 412 nm (at the peak of the Soret band) is 4.8×10^5 M^{-1}cm^{-1}. The extinction in erythrocytes is reduced, due to hypochromism, to $\sim 2.4 \times 10^5$ M^{-1}cm^{-1}. Thus, at 412 nm D_L is ~ 0.06, i.e. T_L is ~ 0.87. The intensity of protein luminescence of an erythrocyte should decrease by 13% due to the microreabsorption at 412 nm. Fig. 5.3 shows the luminescence and ab-sorption spectra for a diluted suspension of erythrocytes. The shape of protein luminescence spectrum is a bit distorted in the range of the heme absorption band at 416 nm. The absorption spectrum of heme in erythrocytes is shifted towards longer

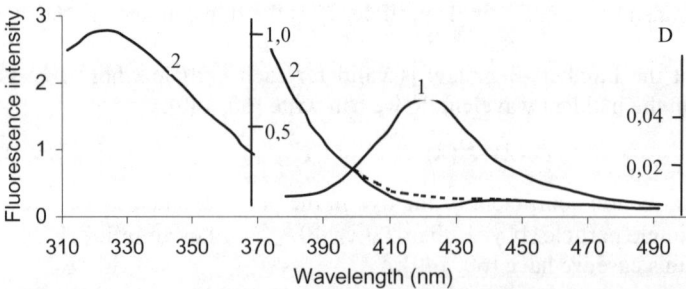

Fig. 5.3. Microabsorption of protein luminescence in erythrocytes (within the Soret band): (1) the absorption spectrum of erythrocyte suspension in 0.15 M NaCl; (2) the spectrum of protein luminescence from erythrocytes; the excitation wavelength is 286 nm. In the range of 370–490 nm the scale is three-fold expanded. The dash courve is the intensity of luminescence in the absence of microreabsorption obtained by interpolation [60, 140]

wavelengths by 4 nm as compared to the spectrum of hemoglobin in solution. The intensity of protein luminescence at 416 nm is decreased by 20% of the intensity that could be observed in the absence of the microreabsorption. The trivial macroscopic reabsorption in the suspension did not exceed 5%. Thus, the experimental value for the microreabsorption is very close to the calculated, value, i.e. about 15% and 13%, respectively.

Similar effects can probably occur in various heterogeneous systems such as visual rods and cones, chloroplasts, biomembranes, colloids, etc.

Consider the situation where cells are stained with some dyes. Cell suspensions and cell organelles having typical dimensions of 10^{-4}–10^{-3} cm are commonly stained with dyes to the final concentration of the dye (if it was distributed over the total volume of solution) of 10^{-4}–10^{-6} M. For instance, let take $d=5\times10^{-4}$ cm, $V=0.5d^3$, $C_R=10^{-5}$ M. If the concentration of particles is 10^{-14} M (the number of particles in Moles per litre of the suspension), and if all the dye molecules are bound to the particles ($\gamma=1$), then substituting these values in equation (5.4) we obtain that C_L is equal to 2.6×10^{-2} M. Let the extinction coefficient of dye molecules in the particles (ε_L) is equal to 5×10^4 M^{-1}cm^{-1}. Substituting the values of ε_L, C_L and d in equation (5.6) we obtain that C_L~0.65, i.e., T_L ~0.22. Thus, the intensity of luminescence from the particles in the suspension with the added dye will be reduced approximately by 78% due to microscreening or microreabsorption.

5.4 Volume Reabsorption of Donor Luminescence

Sensitized fluorescence of the acceptor should be quantitatively estimated to calculate the efficiency of the resonance energy transfer (Chapters 11–14). Thin layers and registration of the emission from the front surface of a cuvette are commonly employed to eliminate reabsorption. Whether this is sufficient for complete elimination of reabsorption is still in question.

Reabsorption of the donor fluorescence by the "acceptor" inside the cuvette may occur not only in the direction of the optical path of the fluorescence collection, but also in all other directions. This "volume reabsorption" should result in an additional increase in the intensity of fluorescence of the acceptor. It is not accompanied by a significant decrease in the donor fluorescence collected from the front surface.

The pathways of the light flows within the cross-section of the excitation beam with the dimensions of h in height and a in length in a cuvette with the optical path d, are shown in Fig. 5.4

If the donor emission propagating in the d direction is collected from the front surface, then its intensity in the presence of acceptor could be written as:

$$F_t=\beta F_0\varphi_\tau(1-T_d)P_d \tag{5.8}$$

where F_0 is the intensity of the exciting light, φ_τ is the quantum yield of donor fluorescence, $T_d=10^{-end}$, $P_d=10^{-kcd}$, where n is the concentration of donor, e and k are the molar extinction coefficients of donor at the excitation wavelength and of

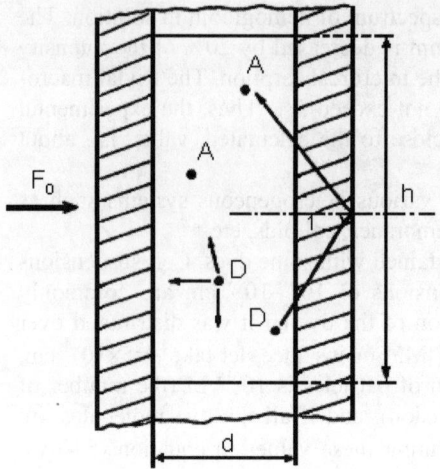

Fig. 5.4. Pathways of light flows in a cuvette. D is the donor, A is the acceptor, i is the angle of total internal reflection, h is the height of the excitation beam. F_0 is the intensity of excitation beam

acceptor in the reabsorption band, β is the coefficient of fluorescence collection. The fluorescence of acceptor, registered from the same surface, is expressed by the following equation:

$$F_r \approx \beta F_0 \varphi_r [\varphi_t(1-T_d)(3-P_d-P_a-P_h)+(1-Q_d)] \tag{5.9}$$

where $P_a=10^{-kca}$, $P_h=10^{-kch}$, $Q_d=10^{-xcd}$, where x is the extinction coefficient of acceptor in the region of donor excitation. As is seen from equation (5.9), while exciting in the region of donor, the acceptor fluorescence is caused by reabsorption in directions of a, h and d as well. For thin layers it is also necessary to take into account the effect of the total internal reflections at the interfaces, namely, at the interfaces between the solution and quartz and between quartz and air. When the incidence angle is above 42 degrees, the fluorescence of the donor reflects to the cuvette, that increases effective optical path. Therefore, it enhances the effect of volume reabsorption (Fig. 5.4). For very thin layers a>>d<<h. Up to 50% of the donor fluorescence which propagates in all directions may return due to total internal reflection. Hence, a coefficient ρ_0 should be introduced into the equation to take into account for the effect of total internal reflection. The volume reabsorption does not disappear in thin layers although the use of these layers eliminates distortions in the emission spectrum of the donor.

For thin layers, where $a>>d<<h$, we have

$$F_r \approx \beta F_0 \varphi_r \varphi_t (1-T_d)(2-P_a-P_h)\rho_0 \tag{5.10}$$

Thus, additional fluorescence intensity of the acceptor could be observed even in thin layers. The intensity of this fluorescence is mainly due to the reabsorption of the donor fluorescence in the directions of a and h. It should be noted that the intensity of donor fluorescence in the direction of a or h is similar to that in the direction of d due to total internal reflection. To quantitatively evaluate the acceptor fluorescence induced by the volume reabsorption in thin layer equation (5.11) should be used:

$$\frac{F_r}{F_t} \approx \varphi_r(2 - P_a - P_h) \tag{5.11}$$

The volume reabsorption that increases the fluorescence of the acceptor becomes more efficient in thin layers. The acceptor fluorescence induced by the volume reabsorption should be decreased with the decrease in the size of the light spot, $a \times h$. In fact, the decrease in $a \times h$ leads to a decrease in the intensity of the fluorescence of acceptor (see below).

5.5 Trypaflavine Activates Rhodamine Fluorescence

Figure 5.5 shows fluorescence spectra of trypaflavine, rhodamine-B and their mixture measured at the front surface of plane quartz cuvette with optical path of 1 mm. When trypaflavine is excited, the intensity of rhodamine fluorescence in

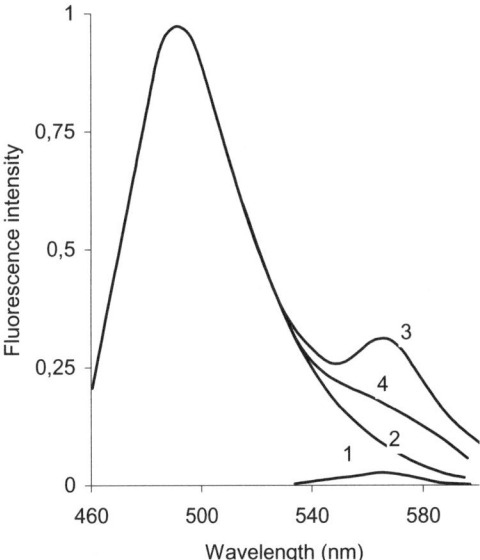

Fig. 5.5. Fluorescence spectra of rhodamine-B (1), trypaflavine (2) and both dyes (3) and (4) in methanol [143]. Rhodamine concentration is 4 µM, trypaflavine concentration is 3 µM. Excitation is at 440 nm. The cross-section of the excitation light beam is: width=6 mm (1–3) and 2 mm (4), height=6 mm (1–3) and 5 mm (4). The optical path equals 1 mm (1–4). "Hitachi-4" spectrofluorimeter

the mixture with trypaflavine becomes significantly higher than that of rhodamine alone.

At concentrations applied in methanol, trypaflavine does not aggregate with rhodamine. The average distance (R) between rhodamine and trypaflavine can be calculated from the following equation:

$$R \approx \sqrt[3]{C^{-1} N^{-1}} \qquad (5.12)$$

where C is the concentration of rhodamine, and N is Avogadro's number. Substituting the values for C and N we obtain $R=700$ Å. At these distances nonradiative transfer is not possible; nor is diffusion-limited collisional quenching, because the lifetime of trypaflavine in methanol is only 4.5 ns. The effects of screening and reabsorption in a 1-mm layer are very small and can be neglected; the absorbance of rhodamine at the maximum of the trypaflavine emission spectrum does not exceed ~0.04 (Fig. 5.6). What is the reason of the increase in the fluorescence intensity of rhodamine in the presence of trypaflavine?

Fig. 5.6 depicts the absorption spectra of trypaflavine, rhodamine and their mixture measured with a "Specord UV-vis" spectrophotometer in a 10-mm cuvette at the same concentrations as in measurements of fluorescence. At 440 nm rhodamine is not actually excited. The reabsorption of trypaflavine fluorescence by rhodamine toward light collection is no greater than 10% per 1 mm of the optical pathway even at the peak of rhodamine absorption. Taking into account that the thickness of the layer toward light collection is about 1.4 mm, the average length of the reabsorption path in this direction is not greater than 0.7 mm. Thus, at the wavelength of 490 nm the reabsorption in the direction of light collection may be neglected. The main role in the increase in the fluorescence intensity of rhodamine belongs only to volume reabsorption, i.e. to the reabsorption of the trypaflavine fluorescence fluxes propagating in others directions.

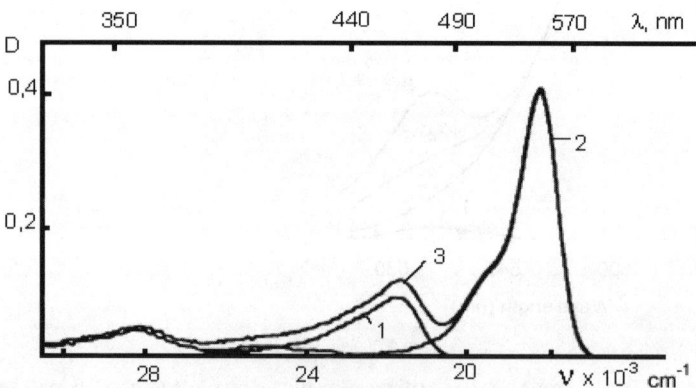

Fig. 5.6. Absorption spectra of trypaflavine (1), rhodamine-B (2) and the mixture of trypaflavine and rhodamine-B (3) in methanol [143]. Rhodamine concentration of is 4 µM, trypaflavine concentration is 3 µM. Optical path equals to 10 mm

Table 5.1 illustrates the effect of a and h on the fluorescence of rhodamine. A decrease in the size of the light spot results in a sharp decrease in the intensity of fluorescence of rhodamine as compared to the intensity of fluorescence of trypaflavine.

The effect of total internal reflection is eliminated by placing the cuvette in a medium with the same refraction index, by increasing the thickness of the cuvette walls, or by blackening the walls (Table 5.1).

Table 5.1. Dependence of the fluorescence of rhodamine on the size of the light spot and on total internal reflection [143]

Light spot size, $a \times h$, [mm]	Conditions	F_t/F_r
6×6	Standard conditions	3.55
2×6	The same	6.6
3×2	The same	~16.0
2×2	The same	~31.0
6×6	Blackening of back wall	5.5
6×6	Increasing the thickness of cuvette wall	7.4
6×6	Placing cuvette in water	7.3

3 µM trypaflavine and 4 µM rhodamine in methanol. Excitation is at 440 nm, trypaflavine emission is at 490 nm, rhodamine emission is at 570 nm, excitation monochromator slits are 4 nm, emission monochromator slits are 8 nm. The fluorescence of rhodamine and trypaflavine is registered at the emission wavelength of 570 nm.

The obtained data explains why abnormally high "sensitised" fluorescence of the acceptor was observed earlier [144, 145] in thin layers of dye solutions. The effect of volume reabsorption is observed even in diluted solutions. Evidently, this is particularly important in the studies of concentrated dye solutions and crystals, and, in particular, in polarization measurements. It should be noted that the first studies that provided evidence in favour of the Forster model of energy transfer were performed with concentrated solutions of trypaflavine and rhodamine [145–149].

The total internal reflection of the donor luminescence can take place at the interface of particle solution, in suspensions of micron-sized particles. This can lead to "micron-volume" reabsorption by acceptor molecules inside each particle [142].

The role of total internal reflection was extensively studied in the measurements of crystal luminescence [150]. The importance of the geometry of the light spot in measurements of luminescence was mentioned in [151]. However, the effects of volume reabsorption and total internal reflection were not quantitatively considered in any of the studies of energy transfer in solutions of trypaflavine and rhodamine [145, 148, 149] or other dyes, polymers or biopolymers.

5.6 Conclusion

Effects of microscreening and microreabsorption can be observed in heterogeneous biological structures. Local absorbances in specific structures can substantially exceed the macroscopic absorbance of the suspension, which results in the decrease in luminescence intensity and distortion in the shapes of excitation or emission spectra. The effect of microreabsorption was demonstrated in the experiment with erythrocytes.

The effect of volume reabsorption in thin layers results in an additional increase in the intensity of luminescence of the acceptor due to the reabsorption of donor luminescence within the volume of the light spot. This effect is experimentally demonstrated using a "classical" Forster pair trypaflavine-rhodamine in methanol. The effect of volume reabsorption is quantitatively estimated and ways of eliminating this effect are recommended.

6 Multipass Cuvettes for Luminescence Spectroscopy

Fluorescence spectroscopy is one of the most sensitive and widely used experimental techniques in physics, chemistry and biology. One of the ways of enhancing sensitivity in fluorescence spectroscopy is increasing the optical path of exciting light in the fluorescent sample. Exciting light passes through the sample two or three times due to reflections from the mirrors placed near the cell filled with the fluorescent solution under study [152–154]. The gain in sensitivity does not exceed a factor of 2 or 3 and also light losses at the interfaces. The mirror-air interface induces an artifact polarization of light. Moreover, due to the reflections at the interfaces the exciting light can get in the path of the emission light, contaminating fluorescence signal.

6.1 Mirror and Total Internal Reflection Cuvettes

To overcome mentioned difficulties, multipass cells providing 4-10-fold gain in the sensitivity and eliminating undesirable reflections and polarizations were designed [155–158].

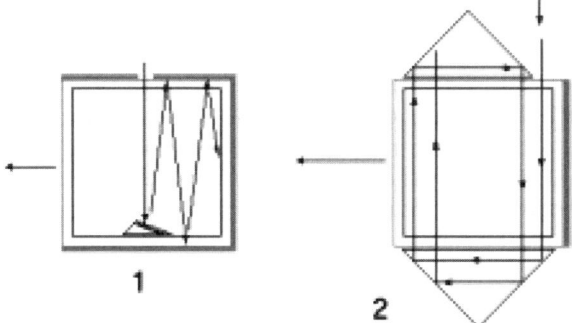

Fig. 6.1. Multipass cuvettes for fluorescence spectroscopy [155–157], viewed from the top: 1) mirror cuvette; 2) total-internal-reflection cuvette. Radiation paths are shown by arrows. Reflective coatings are depicted by hatching

One of these cells is well known in Raman spectroscopy as the "mirror" cell, where mirror-coated walls are used as reflecting elements instead of external concave mirrors [159]. The fluorimetric mirror cell (Fig. 6.1) is a 1-cm quartz cuvette where the outer surface of some walls is coated with aluminium or silver (which is protected by a special coating on the outside). The mirror wall on the front side of the cell has a narrow window, which serves as an entrance for the exciting light. A small narrow prism with reflective faces is used to deflect the incident light beam inside the cuvette. The surface of the prism is protected from solvents by a special coating. The prism can be attached to the rear wall or manufactured together with the wall. The prism provides a small (5–9 degrees) deflection of the incident light from the normal. The exciting light passes through the solution several times due to reflections from the front and rear walls; fluorescence is collected through the transparent side wall; the reflective coating on the opposite side wall improved the efficiency of fluorescence collection. The mirror cell can be used in any spectrofluorimeter without changing its design [155–158].

The "total-internal-reflection" cell (Fig. 6.1) consists of a cuvette with one reflecting side wall and two prisms attached externally to the front and rear walls or manufactured together with these walls. A parallel beam of light enters the cell near the edge of the front face, passes through the solution, experiences two total internal reflection at the interfaces between quartz and air, and returns in the cell, passing it many times. The total-internal-reflection cell can be used most efficiently when the incident light beam is parallel (e.g. laser beam) and the emission light collects from the whole volume of the cuvette. A special holder that allows positioning of the cell relative to the incident beam is used.

The gain in sensitivity (G) from multipass cells (as compared to a common single-pass cell) is given by the following equation [155, 158]:

$$G=(1+RT+R^2T^2+\ldots+R^nT^n)(1+R) \qquad (6.1)$$

where R is the reflectance, T is the transmittance of the solution within the excitation band for one pass, and n is the number of passes. The first multiplier in (6.1) describes the enhancement due to multiple passes of exciting light. The second multiplier accounts for the improvement in fluorescence collection due to the reflection from the reflective side wall. The larger is the transmittance of the solution (i.e., the smaller is the absorbance), the greater is G. The equation is an exact one only if the emission light collects from the whole volume of the cuvette.

The value of R for an aluminium coating in the UV and visible regions is about 0.8–0.9 [155, 159]. Therefore, when T is close to 1 (a weakly-absorbing solution), the value of G for a mirror cell can be as high at least of ~9. The value of R for a silver coating in the UV region is low, but in the visible region it equals 0.95. Therefore, G can approach the value of ~17 after ten passes. The value of R for the total internal reflection cell is close to 1. Using these cells can increase fluorescence intensity by a factor of several tens of times.

However, it is difficult to achieve more than just a few light passes in the 1-cm cell. This limitation results from the beam divergence (excitation beam is not parallel), absorption of light by quartz, and reflection losses. Moreover, if the absorbance of the solution (at the excitation wavelength) is larger than 0.01, the value of

G becomes smaller due to the trivial screening effect. When the absorbance is greater than 0.1, the use of multipass cells is worthless. Moreover, in standard spectrofluorometers the emission light collects mainly from the centre of the cuvette. This limits the gain.

Table 6.1. The gains in sensitivity obtained with different cuvettes [155–157]

Cuvette	$G_{theor.}$	$G_{exper.}$
Conventional 1-cm quartz cuvette	1.0	1.0
Cuvette with two outer concave mirrors	3.3	3.0
Mirror cuvette with aluminium coating	9.0	6.0
Mirror cuvette with silver coating (except for UV)	17.0	8.0
Total internal reflection cuvette	20.0	10.0

Experimental values of G for the 1-cm multipass cuvettes are ranging between 4 and 6 for mirror cells with aluminium coating, between 5 and 8 in visible region for the ones with silver coating, and between 8 and 10 for total-internal-reflection cells when excited by a parallel laser beam. Table 6.1 represents theoretical and experimental values of the gain in the fluorescence intensity obtained with different cuvettes.

6.2 Multipass Cuvettes in Steady–State Measurements

Fig. 6.2 shows fluorescence emission spectra of 10 μM solution of 1-anilino-naphthalene-8-sulfonate (ANS) in ethanol recorded with "Perkin-Elmer MPF 44B'

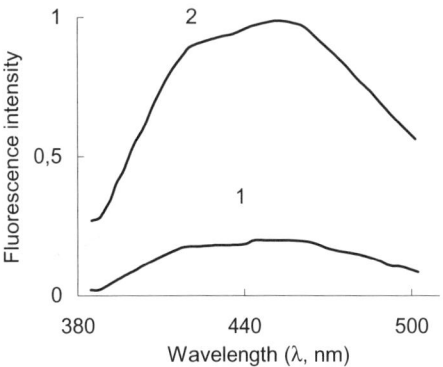

Fig. 6.2. Fluorescence emission spectra of 10 μM ANS in ethanol in a standard cuvette (1) and in the mirror cuvette (2) [156]

spectrofluorometer using a conventional 1-cm cell and a mirror cell. The intensity of fluorescence in the mirror cell is seen to be much greater than in the conventional one, while the shape of the emission spectra is the same. The shape of excitation spectrum does not change if a cuvette with aluminium coating is used (R is independent of the wavelength in the UV and visible regions).

6.3 Multipass Cuvettes in Lifetime Measurements

Since the fluorescence intensity increases in multipass cells, the excited-state lifetime measurements become more accurate. For instance, lifetime measurements using a standard cell filled with ANS solution in ethanol resulted in a distribution of values exceeding 0.5 ns in width (the lifetime equals 7.7 ns; it was measured with the use of the phase-modulation spectrofluorometer "SLM-4800" at the frequency of 30 MHz). The use of multipass cells under the same conditions resulted in a distribution of values no more than 0.1 ns in width.

In multipass cells the exciting light is almost completely absorbed (by the solution and by the cell itself); undesirable artifact reflections at the interfaces are eliminated. Thus, the exciting light is actually prevented from getting into the registering channel. This is particularly important when the emission intensity from the sample is low.

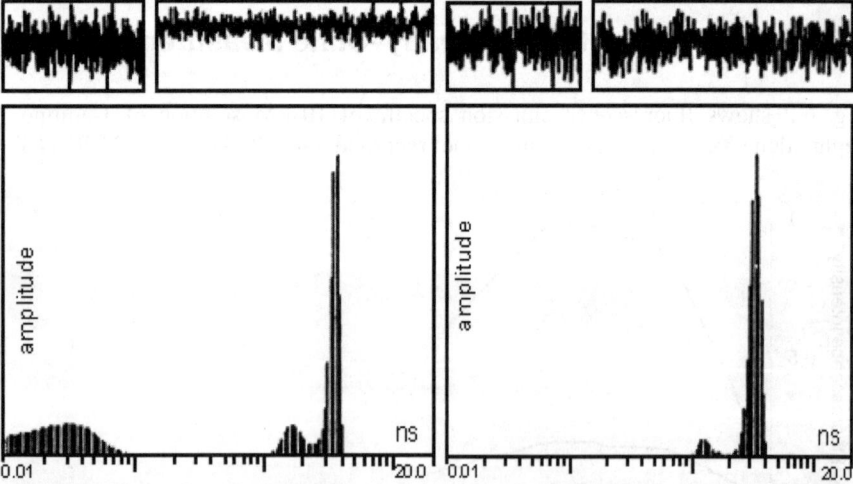

Fig. 6.3a. and **6.3b.** Lifetime distributions of parvalbumin fluorescence obtained using standard cuvettes (left) and multipass cuvettes (right). Cod parvalbumin was dissolved in a 25 mM of Tris-HCl, 1 mM of $CaCl_2$, pH 8.2; the absorbance at 295 nm was less than 0.1; excitation wavelength is 295 nm, emission was detected at 320 nm; the monochromator slits were 5 nm. The measurements were made in collaboration with J. Gallay and M. Vincent at the synchrotron in Orsay

Fig. 6.3 demonstrates the lifetime distributions of tryptophan fluorescence in cod parvalbumin measured in standard cells and multipass cells [158]. Lifetime distributions were measured by the method of correlation spectroscopy [160]. The obtained with standard cells the lifetime artifact, 0.03 ns, is caused by a high noise in a small fluorescence signal and also by penetration of the exciting light into the registering channel. It is prevented if multipass cells are used. In the case of subnanosecond lifetimes the measurements should be made using multipass cells in all channels.

6.4 Other Applications

Multipass cells can also be used in measuring synchronous, resonance and anti-Stokes spectra. The excitation and emission monochromators can be adjusted to similar or identical wavelengths.

The cuvettes may be useful not only for fluorescence analysis, but also for many-fold intensification of laser flash photolysis.

At present, multipass cells for the spectral range of 250–1000 nm are patented and manufactured by scientific and production company in Pushchino (Russia).

6.5 Conclusion

Multipass cuvettes enable a multiple increase in the sensitivity of spectrofluorimetric measurements. The cuvettes can be used for recording emission or excitation spectra and for determining the excited-state lifetime in weakly-absorbing solutions.

7 Division of Tyrosine and Tryptophan Fluorescence Components

7.1 Generally Accepted Approaches

The fluorescence of tryptophan and tyrosine residues is widely used in studies of the structure and dynamics of proteins [18, 22, 25, 67, 161]. Dividing of the tyrosine component from tryptophan component in the fluorescence spectrum of a protein and searching for individual bands of identical aromatic residues located in different microenvironments are an important task, especially in quantitative measurements of the efficiency of energy transfer [60, 162]. The tyrosine and tryptophan components are usually divided using two or three different excitation or emission wavelengths [67, 163]. For example, excitation of tryptophan residues is accomplished in the wavelength range of 285-300 nm, and excitation of tyrosine residues is accomplished at 275 nm [164].

Unfortunately, the tyrosine absorption band extends to nearly 295 nm and can strongly interfere with the selective excitation of tryptophan residues in multi-tyrosine proteins. On the other hand, at 275 nm both tyrosine and tryptophan residues are excited. The emission spectra are wide and they have an overlap in the spectral region between 300 and 320 nm.

If a protein has Trp or Tyr on the surface of the globule, the emission componets can be partially resoluted using glycerol solutions. Fig. 7.1 shows the emission spectra of bovine and porcine phospholipase A2 obtained in a solution containing 90% glycerol and 10% of aqueous acetic buffer (the excitation wavelengths of 275 and 300 nm). Bovine phospholipase contains one tryptophan and seven tyrosines, whereas porcine phospholipase has one tryptophan and eight tyrosines. The tyrosine component in the emission spectrum is revealed at wavelengths shorter than 340 nm. This part is small for bovine enzyme and very large for porcine enzyme. Therefore, only one tyrosine residue of porcine phospholipase is not quenched. Seven tyrosines are quenched by some non-chromophoric groups, but not by the tryptophan residue [243]. The tyrosine component is hardly divided from tryptophan emission in bovine phospholipase. Trp may be excited along at the "red-edge" region of 300 nm. Tyr as well as Trp are both excited at 275 nm, the absorption maximum of Tyr. This is a typical situation for many proteins.

Selection of a number of tryptophan residues located in different microenvion-

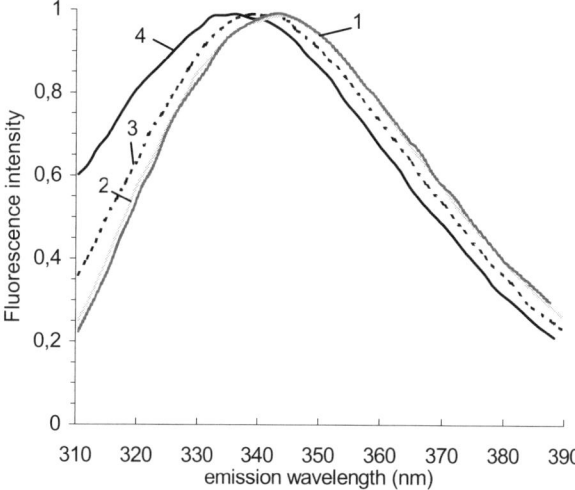

Fig. 7.1. Normalized fluorescence spectra of bovine and porcine phospholipase A2 in a solution containing 90% glycerol and 10% of aqueous acetic buffer. The excitation wavelength is 300 nm {(1) is bovine phospholipase, (2) is porcine phospholipase} and 275 nm {(3) is bovine phospholipase, (4) is porcine phospholipase}; slits are 4 nm, "SLM-4800". The data were obtained in collaboration with J. Gallay and M.Vincent

ments is especially difficult. Emission spectra are wide and significantly overlapped. Among suitable approaches is a time-resolved spectroscopy with phase-suppression of a single lifetime component [165, 166], but the method is not always applicable to proteins with buried tryptophans because of slow relaxation processes after photoexcitation [67, 167].

A quenching of accessible tryptophan residues by iodide, acrylamide and other quenchers is a widely-used steady-state selection method [67, 17, 18, 161]. Using pH-dependences, the individual fluorescence of the two tryptophan residues may be resoluted [168]. Differential spectroscopy techniques are very desirable to be used [163].

While probing the environment of aromatic residues in proteins, high-sensitivity UV resonance Raman spectroscopy is used [169]. For instance, UV resonance Raman spectroscopy and UV resonance Raman saturation spectroscopy (excitation wavelengths at 229 and 238.3 nm) were employed in [170] to examine changes in the environments of the tryptophan and tyrosine residues of human hemoglobin fluoride during structural changes induced by hexaphosphate. The observed average excited-state relaxation rate of the tryptophan residues equals 1/120 ps and is independent of the quaternary structure. The average relaxation rate of tyrosine residues is 1/60 ps.

Two-photon excited fluorescence spectroscopy was applied for tryptophan, tyrosine, phenylalanine in [171, 172]. Two-photon excitation in the visible region is a useful approach to abolish the inner filter effect.

7.2 Synchronous Scanning Method

The synchronous scanning method is of interest in the selection of individual components in protein fluorescence. The method of synchronous spectrofluorimetry, proposed by Lloyd in 1971, was advanced and developed in [173-175]. The method is based on synchronous scanning excitation and emission monochromators with a wavelength shift that enables one to acquire the narrowest and least overlapping bands. Compared to conventional emission spectra [67] and resonance spectra [176], the synchronous spectra do not make clear physical sense, since form and position of the spectral bands change with the wavelength shift between the excitation and emission monochromators. The synchronous spectrum is a superposition of excitation and emission spectra. On the other hand, this method enables to select of individual fluorescence bands in complex mixtures that cannot be obtained by common methods. The method has two advantages over common methods. They are: a) the synchronous spectrum has a much smaller half-width and b) as a rule, it is recorded at the edges of excitation and emission bands. Therefore, even small variations in absorption or/and emission bands affect the synchronous band much more than in the case of a conventional emission spectrum. The spectral interval required for recording the synchronous spectra is considerably shorter [177]. Thus, this method allows to save the time during the experiment. The benefit of the method was demonstrated with mixtures of aromatic hydrocarbons having narrow excitation and emission bands [173-175, 177]. The synchronous method was used for fluorescent probes bound to DNA [178].

7.3 Synchronous Spectra of Tryptophan and Tyrosine

In aqueous solution tryptophan and tyrosine have wide structureless absorption and emission spectra [67]. Therefore, the narrowest synchronous spectra can be obtained by scanning on the edges of the absorption and emission spectra. An optimum wavelength shift between monochromators can be selected, for which the half-width of the synchronous spectrum is 2-4 times less than that of the conventional emission spectrum. It makes it possible to select individual bands. The optimum shift should also meet the following condition: the fluorescence intensity has to be sufficient to record spectra with a high signal/noise ratio.

The first application of the synchronous scanning method to the resolution of tyrosine and tryptophan fluorescence was described in tabulated view by Miller [179]. It was shown that at small $\Delta\lambda$ values the synchronous fluorescence of a tyrosine-tryptophan mixture is characteristic of tyrosine, whereas at large $\Delta\lambda$ values spectra similar to that of tryptophan. Spectral synchronous scanning data on the tyrosine-tryptophan mixture and proteins were published recently [180]. Fig. 7.2 shows conventional emission and synchronous spectra of tyrosine fluorescence in water [180]. The synchronous spectrum of Tyr was obtained by a simultaneous scanning of the excitation wavelength from 260 to 310 nm and the emission wavelength from 280 to 330 nm; the wavelength shift between the monochroma-

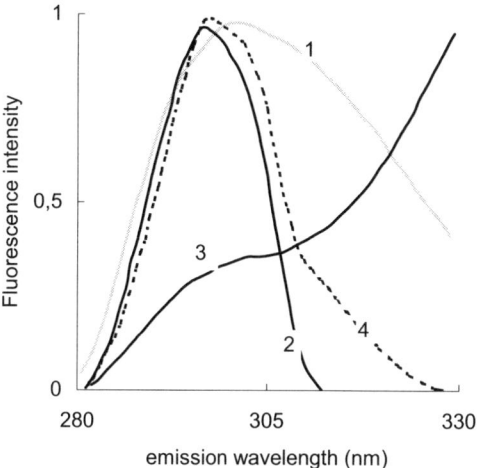

Fig.7.2 Fluorescence spectrum of tyrosine in water excited at 275 nm (1), synchronous spectrum of tyrosine obtained by scanning excitation wavelength from 260 to 310 nm and emission wavelength from 280 to 330 nm (2), fluorescence spectrum of tyrosine and tryptophan mixture (1:1) excited 275 nm (3); and the synchronous spectrum of this mixture obtained by scanning excitation wavelength from 260 to 310 nm and emission wavelength from 280 to 330 nm (4) [180]

tors was 20 nm. The half-width of the synchronous spectrum was 17 nm (half-width of the conventional emission spectrum was 40 nm).

The optimum synchronous fluorescence spectrum of tryptophan in water (not shown) was obtained by scanning excitation wavelength from 275 to 315 nm and emission wavelength from 295 to 335 nm, the wavelength shift between monochromators was 20 nm. The half-width of this synchronous spectrum was 15 nm (half-width of the conventional emission spectrum was 58 nm).

In the case of the tyrosine-tryptophan mixture (1:1) the tyrosine component appears only as a shoulder in the conventional emission fluorescence spectrum, but it becomes dominant in the synchronous spectrum (Fig. 7.2). The difference between the spectra (4) and (2) corresponds to the residual fluorescence of tryptophan. Tryptophan component without contamination from tyrosine can be registered increasing the wavelength shift between monochromators up to 70 nm (not shown).

As is known, bovine serum albuminum contains seventeen tyrosine residues and two tryptophan residues. Tyrosines are strongly quenched in albumin. Therefore, in the conventional emission spectrum the tyrosine component is masked by a very strong tryptophan fluorescence (Fig. 7.3). However, in the synchronous spectrum both tryptophan and tyrosine components are recognisable (Fig. 7.3). Their intensities do not differ so much as in the conventional emission spectrum. Further the-bands could be divided by conventional differential methods.

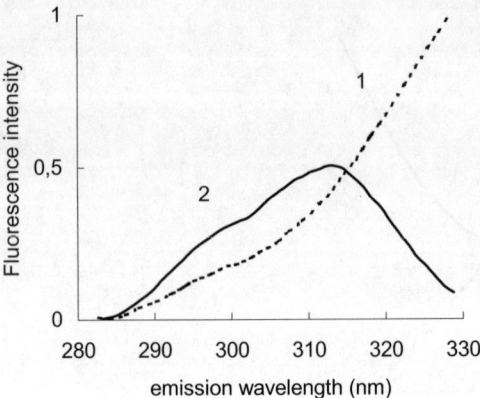

Fig.7.3. Fluorescence spectrum (short-wave part) of bovine serum albumin in a 50 mM phosphate buffer (pH 6.9) excited at 275 nm (1) and the synchronous spectrum of bovine serum albumin obtained by scanning excitation wavelength from 260 to 310 nm and emission wavelength from 280 to 330 nm (2) [180]

7.4 Synchronous Spectra of Different Tryptophan Residues

Individual components from identical aromatic residues located in different microenvironments can be partially divided by conventional methods. For instance, in sperm whale apomyoglobin both tryptophans are neighbours on the N-side of the protein globula [181]. Their normal emission bands are wide and significantly overlapped. It is difficult to divide these bands using conventional methods (see the study of apomyoglobin fluorescence in [181]). The synchronous spectra of tryptophan fluorescence in apomyoglobin are represented in Fig. 7.4. The fluorescence of apomyoglobin results mainly from tryptophan residues, because tyrosine residues are strongly quenched. The absorption spectra of both tryptophan residues do not differ much. Thus, with the use of the synchronous method, two tryptophan components selected mainly by difference in the emission spectra.

The synchronous "blue" and "red" spectra enable to monitor conformational changes in the protein globula. For example, if the value of pH 6.8 decreases up to 5.0, the emission intensity in both components is lowered by approximately equal proportion, close to 35%. The shapes of both spectra do not change in this case (not shown). This proves that the transition to pH 5.0 results in conformational changes in apomyoglobin, in which both tryptophan residues fall in new but identical microenvironments and do not go into the aqueous phase. Therefore, some quenching group moves closer to the tryptophans. At pH 3.3 both the intensities and shapes of the synchronous spectra are changed (Fig. 7.4). The "blue" band

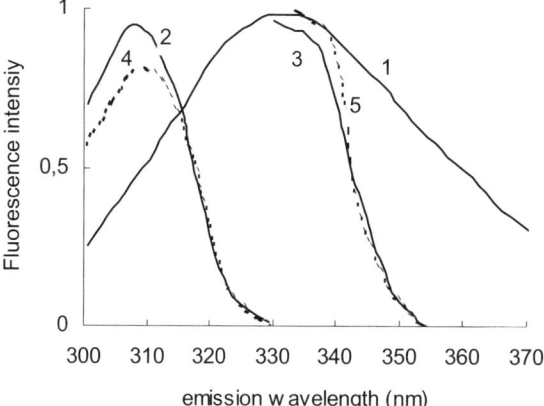

Fig. 7.4. Fluorescence spectra of sperm whale apomyoglobin in a 50 mM phosphate buffer (pH 6.8): emission spectrum excited at 285 nm (1); synchronous scanning of excitation wavelength from 285 to 315 nm and emission wavelength from 300 to 330 nm (shift between monochromators was 15 nm) (2); synchronous scanning of excitation wavelength from 285 to 315 nm and emission wavelength from 330 to 360 nm (shift between monochromators was 45 nm) (3), and synchronous spectra at conditions like (2, 3) but pH was 3.3 (4, 5) [180]

becomes weaker and shifts towards the long wavelengths. However, the "red" band gets stronger and sharper (as compared with the band at pH 6.8). This means that the protein changes to a new conformation in which both tryptophan residues are shifted into a more polar environment.

The above data show that division of individual synchronous bands of tyrosine and tryptophan in solution or in proteins and separate detection of these residues in different conditions can be achieved by varying the wavelength shift between monochromators and the ranges of scanning.

7.5 Conclusion

The method of synchronous scanning proves useful in separating fluorescence components from tyrosine and tryptophan in solution and in proteins. Selecting an appropriate wavelength shift between excitation and emission monochromators makes possible separate detection of "blue" and "red" components of protein fluorescence and registration of conformational changes.

8 Spectral Heterogeneity of Tryptophan Emission

Spectroscopic methods are used in the study of conformational behaviour of proteins on different time scales. Based on the depolarization of the tryptophan fluorescence in proteins and on the quenching of internal tryptophan residues by substances of low molecular weight, some authors postulate the existence of considerable spontaneous thermal mobility of the protein structure on the nanosecond time scale [67, 182]. However, photoexcitation of tryptophan residues may induce non-spontaneous mobility [60, 167, 183, 184].

A liquid solution of a chromophore as a rule has a homogeneous fluorescence spectrum: the emission spectrum is independent of the excitation wavelength, the decay curve is single-exponential, the lifetime does not change along the emission spectrum, etc. Fluorescence spectra of proteins in aqueous solution are generally heterogeneous.

8.1 Variation of Fluorescence Polarization along Tryptophan Emission Spectrum

Many years ago it was found out that the polarization of fluorescence (P) of some proteins are decreased in the long-wavelength region of the emission spectrum [15]. To elucidate the reason, several hypotheses were proposed, among them: (a) the "duality" of tryptophan fluorescence [185] induced by the known [186] duality of absorption by ^1L_a and ^1L_b oscillators [22, 187]; (b) the variation in the efficiency of energy migration between tryptophan residues of different emission spectra [188]; (c) the rotational heterogeneity of emission centres, i.e. different rotational mobility of various tryptophan residues in protein macromolecules [15, 18]; (d) the increase in the angle between absorption and emission oscillators with the emission wavelength [189]; (e) some researchers assumed that P is changed along the fluorescence spectrum due to penetration of excitation light in the registration channel [7]. Actually, every listed reason could be essential. Here we examined this issue in detail [167, 183].

Unlike the above studies, our experiments involved lifetime measurements along the emission wavelengths. The lifetimes were detected by phase-modulation technique. The phase and modulation lifetimes (τ_p and τ_m) were calculated from the phase angle (φ) and demodulation factor (m) [67]:

$$\tau_p = \omega^{-1} \tan\varphi \quad \text{and} \quad \tau_m = \omega^{-1} (m^{-2}-1)^{1/2} \tag{8.1}$$

where ω is the modulation frequency multiplied by 2π.

Each polarization point was calculated using measurements the four intensities at four positions of the two polarising prisms: (a) $0°_{ex}$ and $0°_{em}$, (b) $0°_{ex}$ and $90°_{em}$, (c) $90°_{ex}$ and $0°_{em}$, (d) $90°_{ex}$ and $90°_{em}$. This methodology allows to remove an artefact polarising effect of monochromators, since the polarization is calculated by expression:

$$P=(a/c-b/d)/(a/c+b/d) \tag{8.2}$$

Fig. 8.1 shows typical spectral curves (for hyaluronidase). As is seen the lifetime is increased and polarization is decreased with the emission wavelength. Similar results were obtained for all studied proteins. Table 8.1 represents data on the lifetime, polarization, quantum yield and peaks of emission spectra for some proteins, polypeptide glucagon and DL-tryptophan in aqueous buffers. All proteins including single-tryptophan proteins demonstrate an increase in the lifetime and a decrease in polarization in the long-wavelength region. A similar pattern is observed with the solution of DL-tryptophan in glycerol. The fluorescence of glucagon and DL-tryptophan in water is almost completely depolarised and shows no changes in the lifetime along the spectrum.

Mazurenko et al. [190] were the first who tried to explain the variation in P along the emission spectrum of fluorophores (dyes) in viscous solutions using the known Levshin-Perrin equation. They calculated the variation in P on the basis of lifetimes along the emission spectra of dyes (but did not prove it experimentally). Let us use a similar approach to the tryptophan fluorescence. The Levshin-Perrin equation may be written as:

$$1/P - 1/3 = (1/P_0 - 1/3)(1 + RT\tau/\eta V) \tag{8.3}$$

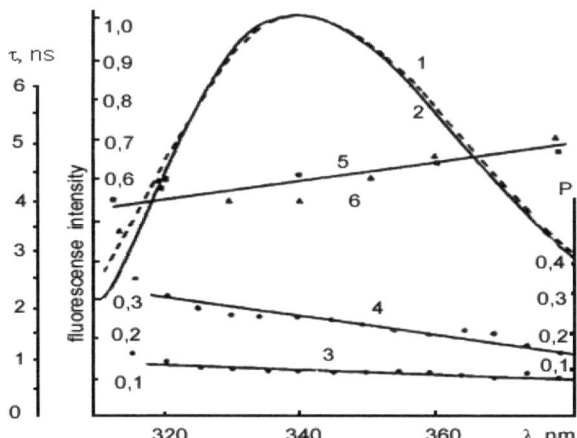

Fig. 8.1. Fluorescence of hyaluronidase. Emission spectra (1) and (2), emission polarization spectra (3) and (4), and variation in lifetime with emission wavelength (5) and (6). The excitation wavelength is 280 nm for 1, 3 and 5; and 300 nm for 2, 4 and 6. The data [183] were obtained with the spectrofluorimeter "SLM-4800"

where P_0 is the limiting polarization, η is the viscosity, T is the absolute temperature, R is the universal gas constant, and V is the volume of a particle. The equation establishes a relationship between the lifetime and the polarization P. The greater is the lifetime, the greater is the angle of rotation of a chromophore, and, consequently, the smaller is P. The following equation relates P to the rotation angle θ [10, 67]:

$$1/P - 1/3 = (1/P_0 - 1/3)[2/(3\cos^2\theta - 1)] \qquad (8.4)$$

For the majority of proteins in solution and for DL-tryptophan in glycerol the value of P is about 0.3 in the short wavelength region of the spectrum, i.e. at 320 nm (Table 8.1). A θ value of about 25 degrees can be obtained assuming that the value of P_0 is between 0.4 and 0.5. This is the existence of rotational mobility of tryptophans in the proteins.

If the value of $P_0=0.5$, $\eta=10$ poise, $R=8.3$ J mol^{-1}K^{-1}, $T=294$ K, $V\sim30$ Å^3 and $\tau=4.2$ ns for DL-tryptophan in glycerol, we obtain that P is equal to ~0.33. The calculated value of P is close to the experimental value of this quantity for DL-tryptophan in glycerol measured with the excitation wavelength of 300 nm and the emission wavelength of 320 nm (Table 8.1 and Fig. 8.2). If $\tau=5.4$ ns and the values of the other parameters are the same, the value of P equals to ~0 3, which is close to the experimental value obtained with the excitation wavelength at 300 nm and the emission wavelength of 380 nm (Table 8.1 and Fig. 8.2).

Thus, P may change along the fluorescence spectrum of DL-tryptophan in glycerol due to variations in the lifetime. According to the Levshin-Perrin equation, the molecules with a longer lifetime rotate through a greater angle, that re-

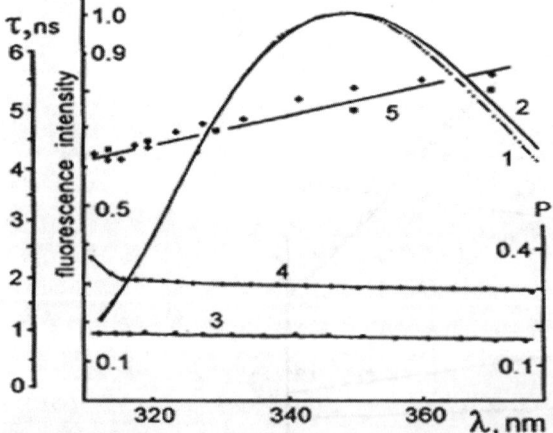

Fig. 8.2. Fluorescence of DL-tryptophan in 95.5% glycerol: fluorescence emission spectra obtained with the excitation wavelengths at 280 (1) and 300 nm (2), fluorescence polarization spectra obtained with the excitation wavelengths at 280 (3) and 300 nm (4); the lifetime measured with the excitation wavelength at 290 nm (5). Concentration of DL-tryptophan is 65 µM. [167]

sults in additional depolarization in the long-wavelength region. The rotational mobility of tryptophan is the main reason of the change in the value of P along the emission spectrum of DL-tryptophan in glycerol.

Three types of mobility could be observed in proteins. They are (a) mobility of tryptophan residues per se; (b) mobility of some protein segments containing tryptophan residues; (c) mobility of the protein globule as a whole (rotational dif-

Table 8.1. Peak emission wavelengths, quantum yields, polarizations and the lifetimes of the excited state of tryptophan in proteins, glucagon and DL-tryptophan

	λ [nm]	Q_f	P		τ [ns]		n
			320 [nm]	380 [nm]	320 [nm]	380 [nm]	
DL-tryptophan in water	347	0.20	0.01	0.00	3.1	3.2	1
DL-tryptophan in glycerol	343	0.34	0.33	0.29	4.2	5.4	1
Glucagon	345	0.07	-	0.04	3.0	3.1	1
Subtilisin	345	0.04	-	0.04	3.4	4.2	1
Peroxidase	-	0.03	0.36	0.23	2.6	4.1	1
Bovine albumin	342	0.32	0.32	0.28	5.5	6.3	2
Phosphoglycerate kinase	340	0.05	0.32	0.13	1.9	3,4	2
Trypsin	328	0.09	0.30	0.22	3.5	4.4	4
Alcohol dehydrogenase	328	0.32	0.34	0.30	4.0	5.1	4
Pepsin	342	0.38	0.22	0.21	6.4	7.2	6
Lysozyme	332	0.07	0.30	0.20	1.6	2.4	6
Chymotrypsinogen	327	0.09	0.23	0.20	2.0	2.7	8
Lactate dehydrogenase	339	0.46	0.33	0.31	6.5	8.4	24
Hyaluronidase	-	-	0.30	0.16	3.9	4.7	-
Hexokinase	-	-	0.32	0.24	3.1	4.2	-
Amylase	336	0.28	0.33	0.27	4.3	5.7	-
Sarcoplasmic Ca^{2+}-ATPase	328	0.30	0.34	0.25	4.2	5.9	19

Experimental conditions: the excitation wavelength was 295 nm; for polarization measurements the excitation wavelength was 300 nm. The polarization and the average lifetime are registered with an "SLM-4800". The peaks of the emission spectra and fluorescence quantum yield, Q_f, were registered with a "Perkin-Elmer MPP-44B"; n is the number of tryptophan residues. The temperature is 20°C. [167].

fusion). All these mobilities may lead to the fluorescence depolarization.

As for large globules, like Ca^{2+}-ATPase of sarcoplasmic reticulum, and large proteins as peroxidase or lactate dehydrogenase, the rotational mobility of the protein globule does not result in depolarization. Substituting in the Levshin-Perrin equation the values of $V \geq 80\,000$ Å^3, $\eta \geq 0.01$ poise and $\tau \leq 6$ ns (the values of the other parameters are the same as for DL-tryptophan in glycerol), we obtain P close to P_0, i.e. the rotation of the protein globule as a whole can be neglected. Therefore, only the rotational mobility of tryptophan residues and tryptophan- containing segments is responsible for depolarization [67, 161, 191, 192]. For small proteins and glucagon polypeptide the rotation of the globule is hardly negligible.

The values of P along the fluorescence spectra of proteins are probably varied because of the rotational mobility as in the case of DL-tryptophan in glycerol. Actually, some other factors may also effect on the change in P along the emission spectrum. For instance, Fig. 8.3 demonstrates spectral curves obtained with bovine serum albumin. The P values in the emission wavelength region from 325 to 304 nm is increased mainly due to the contribution from tyrosine residues, which are rather numerous in albumin. High polarization in the region from 304 to 300 nm results from the penetration of excitation light into the registration channel it and can be eliminated by using shorter excitation wavelengths.

The penetration of excitation light in the registration channel is also observed in the case of DL-tryptophan in glycerol, where the values of P are increased between 310 and 300 nm (Fig. 8.2). This increase in P can be eliminated by using the excitation wavelength of 280 nm. P is changed along the emission spectrum of proteins and DL-tryptophan in the wavelength range between 320 nm and 380 nm.

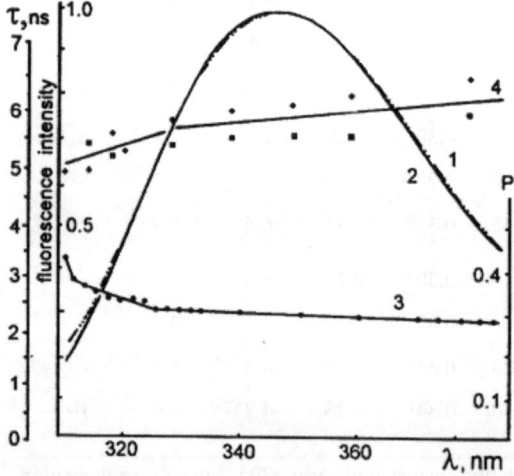

Fig. 8.3. Fluorescence of bovine serum albumin. Emission spectra measured using excitation wavelengths at 290 nm (1) and 300 nm (2); polarization spectrum obtained with the excitation wavelength of 300 nm (3); fluorescence lifetimes measured using excitation at 290 nm (crosses) and 300 nm (squares) (4). [167]

8.1 Variation of Fluorescence Polarization

Table 8.2. Fluorescence lifetimes (ns) of tryptophan in proteins, glucagon, and DL-tryptophan measured at varying excitation wavelengths (at constant emission wavelength of 350 nm)

Wavelengths:	Lifetime [ns] at the excitation		
	280 nm	290 nm	300 nm
DL-tryptophan in water	3.3	3.3	3.4
DL-tryptophan in glycerol	5.0	5.0	5.0
Bovine albumin	6.2	6.5	6.0
Lysozyme	2.1	2.1	2.1
Chymotrypsinogen A	2.3	2.3	2.2
Hexokinase	4.2	4.2	4.3
Glucagon	3.1	3.0	3.2
Sarcoplasmic calcium ATPase	5.9	5.8	5.7

These changes do not related with tyrosine or penetration of excitation light. They are solely due to the properties of tryptophan fluorescence *per se*. For DL-tryptophan in glycerol energy migration is hardly possible because the distances between fluorophores are too large at low concentrations. Moreover, migration is impossible at red-edge excitation at 300 nm. The lifetime for DL-tryptophan in water is almost constant along the emission spectrum (Fig. 8.4). Therefore, duality of fluorescence and variation in the angle between oscillators are not essential.

No heterogeneity in glycerol was found in the excitation spectrum: the spectra monitored by the emission at 320 nm and 380 nm coincide and the value of the lifetime does not change along the excitation spectrum (Fig. 8.5). Similar data are obtained for proteins (Table 8.2).

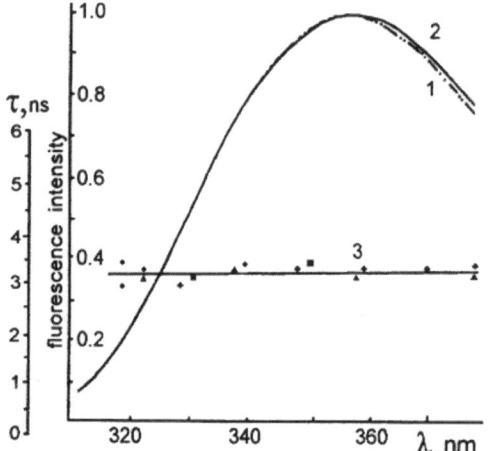

Fig. 8.4. Fluorescence of DL-tryptophan in water. Emission spectra obtained using excitation wavelengths at 280 or 290 nm (1) and 300 nm (2), fluorescence lifetime measured with the excitation at 280 nm (triangles), 290 nm (crosses) and 300 nm (squares) (3). Concentration of DL-tryptophan is 27 µM

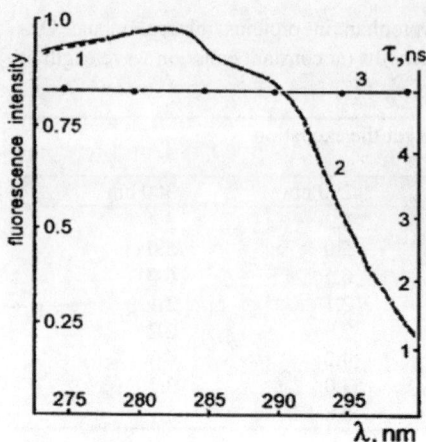

Fig. 8.5. Excitation spectra (1) and (2) and the lifetime (3) for DL-tryptophan in glycerol. The emission was detected at wavelength of 320 nm (1), 380 nm (2) and 350 nm (3). Concentration of DL-tryptophan was 65 µM. [167]

8.2 Fluorescence Lifetime Variations

What is the original course why the lifetime is increased with the emission wavelength for tryptophan in glycerol and proteins?

Lifetime variation along the fluorescence spectrum of dyes in viscous media was described in [193, 194]. The effects atypical of liquid solutions, as the dependence of the emission peak, quantum yield, and the fluorescence lifetime on the excitation wavelength, a time-dependent shift of the time-resolved emission spectrum on the nanosecond time scale, nonexponential decay curve, etc can be observed in viscous solutions and in biopolymer systems [60, 67, 193–195]. A combination of all these phenomena is known as the "spectral heterogeneity". One of the major reasons for the spectral heterogeneity of emission is the orientational and vibrational relaxation of the near environment of the chromophore. This phenomenon is usually described by the model of universal molecular interactions [67, 85, 193, 194].

Two lifetimes were observed at different emission wavelengths in gramicidin-A [196] (gramicidin A has four tryptophan residues in 9, 11, 13 and 15 positions). The major component (pre-exponential factor, ~0.9) has a lifetime of 0.5–1.6 ns and the minor component has a lifetime of 2.6–2.8 ns. Both lifetime components increase with the emission wavelength from 330 to 380 nm. The authors state that the average lifetime increases by 62% as the wavelength is varied from 330 to 380 nm (the lifetime in Figure in [196] changes really from 1 to 3 ns, i.e. the increase is greater than 62%). The steady-state polarization slightly decreases across the emission spectrum. This can be explained by the rotational mobility of tryptophan chromophores. When gramicidin is incorporated into dioleoyl-glycero-3-

8.2 Fluorescence Lifetime Variations

phosphocholine membranes, the emission spectrum of the tryptophans in the gramicidin channel depends upon the excitation wavelength known as the red-edge excitation effect. As the excitation wavelength changes from 280 to 300 nm, the peak of emission shifts by about 2 nm, and as the excitation wavelength changes from 300 to 310 nm, the peak of emission shifts by 4 nm. Unfortunately, the effects of Raman scattering and penetration of exciting light in the registration channel were not taken into account in these experiments.

Spectral emission heterogeneity was also described for other natural chromophores. For instance, emission spectra of porphyrin in metal-free and Zn cytochrome-C and in metal-free mesoporphyrin derivatives of horseradish peroxidases-A and -C, leghemoglobin, and myoglobin were analysed [197]. At a low temperature (4.2 K) the emission spectra depended upon the excitation wavelength. A narrow-band excitation in the short-wavelength region of the 0-1 and 0-0 bands led to structureless emission, whereas excitation in the long-wavelength region formed quasi-discrete spectra.

The model of universal molecular interactions [67, 85, 193, 194] may be described as follows. In viscous media, where the relaxation time of the environment is comparable with the lifetime of a photoexcited chromophore, $\tau_{rel} \sim \tau$, emission starts from the "hot" nonrelaxed states during vibrational and orientational relaxation of a solvent shall near the photoexcited chromophore. Thus, time-resolved fluorescence spectra registered on the nanosecond scale shift from the "blue" to the "red" region [15, 60, 194, 198].

The increase in the average lifetime of tryptophan with the emission wavelength in viscous media indicates that the time-resolved emission band shifts from shorter to longer wavelengths on the nanosecond time scale. The time dependence of the fluorescence of Trp and indole in ethylene glycol/aqueous solution was studied in [199]. Time resolved spectra as a function of temperature display a "red" shift in the emission band on the nanosecond time-scale after photoexcitation.

J.Lakowicz and H.Cherek [200] observed a variation in the lifetime across the fluorescence spectrum of N-acetyl-L-tryptophanamide (NATA) in viscous propylene glycol. The effect was observed only within a certain temperature range and vanished when the temperature increased up to 63°C. In the case of 1,3-dimethylindole in butanol an increase in the lifetime with the wavelength change from 320 nm to 380 nm was registered only within the temperature range between 190 and 220 K, where $\tau_{rel} \sim \tau$; however, the effect was not observed at $T=85$ K and at $T=280$ K because in these cases $\tau_{rel} >> \tau$ or $\tau_{rel} << \tau$, respectively [201].

Thus, at least in the case of DL-tryptophan in glycerol, it could be concluded that the variation in the lifetime along the emission spectrum is a result of spectral heterogeneity resulted from the nanosecond relaxation processes in solvent. Solvent relaxation in water takes a few tens of picoseconds [85, 194, 198]. Therefore, the lifetime along the emission spectrum of DL-tryptophan in water should change only a bit.

In glucagon, a polypeptide with a single chain of 29 amino acids, the tryptophan residue is exposed to water [202]. The emission properties of this tryptophan residue are similar to those of DL-tryptophan in water (Table 8.1). At a neutral pH glucagon is a random coil [202, 203]. There are a great number of glucagon conformers [202, 203]. Our data are correlated with those of Cockle and Szabo [202], who estimated the lifetime of glucagon to be about 3.2 ns at 320 nm and 3.4 ns at 370 nm, as determined by kinetic decay. Thus, the presence of conformers does not necessarily lead to a variation in the lifetime along the emission spectra of proteins.

Unlike glucagon, most of the proteins in Table 8.1 have inner tryptophan residues inaccessible to water. Thus, the mobility of the tryptophan chromophore is strongly restricted and here $\tau_{rel} \sim \tau$. In all proteins the lifetime is increased with the wavelength. No decrease in the lifetime is observed in any protein (Table 8.1). Therefore, the lifetime is varied along the emission spectrum due to relaxation processes. For multi-tryptophan proteins the heterogeneity of centres is the additional reason for lifetime variation. For the single-tryptophan protein, the relaxation processes are the main reason. Harris and Hudson demonstrated the increase in the lifetime with the emission wavelength in single-tryptophan proteins obtained by methods of genetic engineering [204, 205].

The relaxation processes are also occurred in course of fluorescence of the probes bound in the hydrophobic pockets in proteins (Chapter 15).

8.3 Photoinduced Conformational Mobility of Proteins

With the results of our experiments we may explain a serious discrepancy between the theoretical values of tryptophan rotational mobility calculated on the basis of a spontaneous mobility [188] and the experimental data on high rotational mobility of photoexcited tryptophans in the fluorescence measurements [67, 206, 207]. The mobility of the fluorescent proteins is not only spontaneous, but also photoinduced, nonequilibrium. The induced mobility significantly exceeds regular thermal spontaneous mobility when a photoexcited chromophore is buried in the protein matrix. Unfortunately, many investigators erroneously interpreted the dynamics of macromolecules observed in fluorescence experiments as equilibrium dynamics. This interpretation is valid only when $\tau_{rel} \ll \tau$. For proteins $\tau_{rel} \ll \tau$ when tryptophan residues are exposed to the aqueous phase. The measurements of fluorescence of the inner tryptophan residues provide us the information not only on equilibrium mobility, but also on photoinduced dynamics.

After photoexcitation of tryptophan in glycerol and other viscous solvents, the solvent molecules are involved into the nanosecond-scale orientational and vibrational relaxation processes. In proteins the tryptophan residues are surrounded by a few groups fixed on the rigid polypeptide chain. Therefore, photoexcitation induces the relaxation not only in the nearest neighbours but also in the entire protein macromolecule. Thus, photoexcitation induces structural mobility, i.e. the conformational relaxation of a protein. This photoconformational relaxation occurs on the nanosecond time scale. The energy loss is defined as:

$$Q = h\nu_{ex} - h\nu_{em} \tag{8.5}$$

8.3 Photoiduced Conformation Mobility of Proteins 71

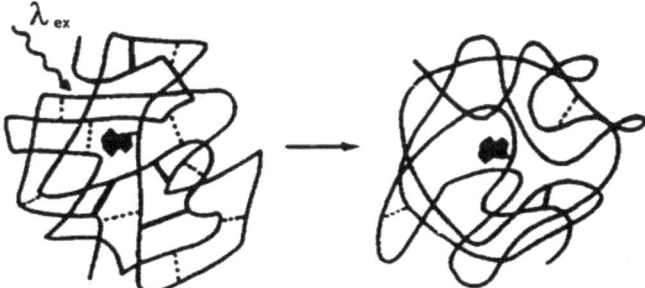

Fig.8.6. Photoinactivation of enzyme induced by photoconformational relaxation [167]

If λ_{ex}=290 nm and λ_{em}=340 nm, the energy loss is 60.6 kJ mol^{-1}, that is sufficient to break four hydrogen bonds. In the absence of emission, Q is ~410 kJ mol^{-1}. Therefore, for most proteins with a low fluorescence quantum yield (Table 8.1), UV-illumination should result in denaturation. For enzymes denaturation is equivalent to inactivation (Fig. 8.6).

According to the traditional viewpoint the cause of photoinactivation of enzymes is photolysis, i.e. a photodestruction of tryptophan residues. The photophysics and photochemistry of tryptophan and its simple derivatives are reviewed in [209]. However, the photoinactivation proceeds always much more rapidly than the photolysis of tryptophan residues [208, 210, 211]. For instance, the results obtained with hexokinase in [208] are shown in Fig. 8.7. Using fluorescence techniques, it was demonstrated that irradiation of proteins (trypsin and trypsinogen) in the UV-region resulted in the "unfolding" of the protein globule and a shift of the steady-state fluorescence spectrum to the longer wavelengths of tryptophan in an aqueous environment [212].

In biomembranes the processes of vibronic relaxation and internal conversion may have a biological function because these processes result in an instant "heating" of the chromophore microenvironment. This heating may lead to an increase

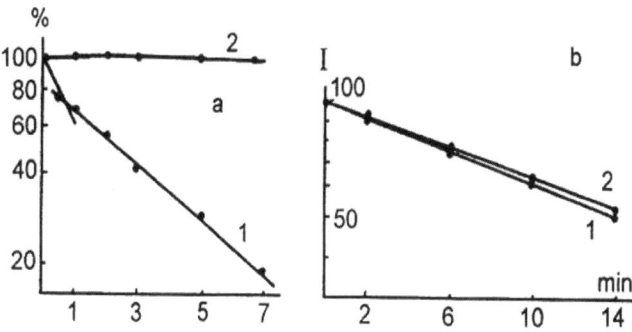

Fig. 8.7. Photoinactivation (a) and photolysis (b) of free hexokinase (1) and hexokinase bound to insulin (2) [208]. Insulin acts as a stabilizer for hexokinase

in the rotational and lateral mobility of the protein globule and also to the mobility of lipids. Photoconformational relaxation can lead to a change in the accessibility of the active centres to ligands or substrates, desorption of a product from an enzyme, etc. (Chapter 17). This may result in an enhancement of enzymatic activity (Chapters 17–19). We may assume that these effects occur in the electron transfer in the process of photosynthesis, in the process of vision, etc.

Using fluorescence quenching, a conformational change accompanying the phototransformation of phytochrome-A was observed [213]. Phytochrome-A exists in two interconvertible photoisomers, the first one being capable of absorbing red light and the second one being physiologically active. Photoisomerization of the chromophore induces alterations in the secondary structure as well as topographic rearrangement. Changes in the protein secondary structure seem to be confined to the N-terminal region, where a photoreversible increase in the α-helix content occurs. The regions around four of the 10 tryptophans are conformationally photoresponsive areas in phytochrome-A. Topographic changes associated with phototransformation are not confined to the 58-kDa domain, and they involve most if not all of the region from Trp-365 to Trp-787.

The possibility of photoinduced relaxation dynamics on the nanosecond time scale should also be taken into account while measuring the depth of penetration of various quenchers into the protein globule. Many researchers studied the accessibility of tryptophan residues to oxygen, acrylamide, etc. However, the photoinduced mobility capable of increasing the steric accessibility of tryptophans to the quenchers was not considered.

8.4. Phosphorescence of Proteins

Phosphorescence in the deoxygenated suspensions of human erythrocyte membranes, yeast, and animal cells at room temperatures was detected by Mazul et al. [214]. Tryptophan nature of the phosphorescence was established. The maxima of emission were at ~407, ~435 and ~463 nm. The lifetimes were of 0.05–3.0 seconds. It was shown that the phosphorescence parameters are sensitive to conformational transitions.

Phosphorescence decay of some proteins is nonexponential. For instance, nonexponential phosphorescence decay was observed for horse liver alcohol dehydrogenase and alkaline phosphatase at room temperature [215]. The authors concluded that nonexponential behaviour could be caused by ground-state conformers and excited state reactions. A mathematical model of phosphorescence decay with heterogeneous distribution of quencher was developed in [216].

Phosphorescence changes in horse liver alcohol dehydrogenase induced by NADH and NAD+ at different temperatures were measured in [217]. The phosphorescence emission spectrum in rabbit muscle aldolase at room temperature before and after the NADH binding was obtained in [218].

The indole phosphorescence spectrum was calculated using the quantum consistent force field method [187].

8.5 Conclusion

Photoexcitation of Trp residues in proteins results in an emission heterogeneity mainly caused by relaxation processes. This effect is revealed as an increase in the average lifetime with the emission wavelength. According to the Levshin-Perrin equation, this leads to depolarization of the «red» emission during tryptophan rotational mobility.

In a viscous media such as a protein globule, the mobility of photoexcited tryptophan in the nanosecond time scale is not only spontaneous, but also an induced, nonequilibrium.

9 Discrete Emission States in Photoexcited Tryptophan Complexes

The nonexponential decay of fluorescence of single-tryptophan proteins and tryptophan in polar solutions is an interesting phenomenon in photobiophysics. The nonexponential decay of tryptophan fluorescence in solutions was studied and discussed in [184, 219–223].

A widely accepted point of view is the ground-state conformer model can explain the phenomenon [202, 220–225]. According to the model, proportions between the excited-state populations are determined by the distribution of the ground-state rotamers. The rate of rotamer interconversion should be slower than the rate of the excited-state relaxation. This model appears to adequately interpret the data obtained with rotationally constrained tryptophan derivatives. However, some data do not agree with the model. For example, the model is not suitable to describe a nonexponential decay of some compounds of a single conformation. Moreover, for a jet-cooled tryptamine (tryptamine has not any acceptor group on the side chain) the differences among the lifetimes of differing conformers are small, ~10% [226].

A monomeric α-helix forming peptide and a dimeric coiled-coil forming peptide containing a central tryptophan residue were synthesized and studied by Szabo and coworkers [220]. The peak of emission was independent of the fractional α-helical content. The decay times were independent of both α-helix content and the emission wavelength. Amplitudes of the decay time components were dependent from the α-helix content. Three decay times were associated with the three ground-state rotamers of the tryptophan residue.

Heterogeneity of emission centers due to ground-state conformers is feasible. However, also a number of the excited-state conformers (photoconformers) could arise at photoexcitation. This idea was mentioned by Fleming and coworkers [184]. They discovered that zwitterions of tryptophan, tryptophenylalanine and alanyltryptophan show a nonexponential fluorescence decay, which can be analyzed as two exponentials. The results were interpreted in terms of two kinds of trapped conformers in the excited state that interconvert no quicker than the time scale of the fluorescence. Unfortunately, the values of both lifetime components were defined with too large error (more precise measurements were carried out by Szabo and Rayner [225]). Recently it was shown the conformer interconversion during the excited state of constrained tryptophan derivatives [529].

In various proteins the average lifetime is increased with the emission wavelength and in glycerol solution of tryptophan as well [60, 167, 183]. To explain this and other facts, the relaxation model was proposed and developed [67, 167, 199, 201, 227, 228]. This model enables to calculate interactions between a fluorophore and solvent. The relaxation model is based on numerous data on the non-exponential fluorescence decay of tryptophan, indole, its derivatives, and probes in viscous media. According to this model, the relaxation time of a viscous solvent is comparable to the fluorophore lifetime; therefore, the fluorophore microenvironment relaxes in the course of emission that results in the spectral heterogeneity of fluorescence. However, the emission heterogeneity of Trp and other indole derivatives in polar solvents cannot be explained completely within the traditional relaxation model (see below).

9.1 Time-Resolved Spectroscopy of Tryptophan Fluorescence

The fluorescence decays in the emission spectrum of tryptophan (Trp) and N-acetyl-tryptophanamide (NATA) in water, ethanol, and glycerol were measured in [219] using time-correlated single-photon counting technique [160] and analyzed by the method of maximum entropy [229]. Last-squares analysis of time-resolved decays and methods of calculation of the goodness fit (χ^2) are described in [67,229].

Figure 9.1 shows a typical evolution of the fluorescence lifetime distribution during maximum entropy calculations.

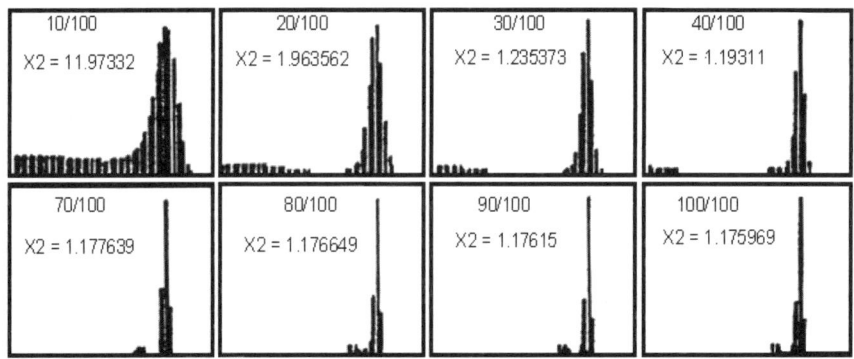

Fig. 9.1. Evolution of the fluorescence lifetime distribution during maximum entropy calculations. X-axis is the lifetime in logarithmic time-scale, Y-axis is the lifetime amplitude, χ^2 is the fit. Analysis of decay as a sum of 100 exponentials was performed by M.Vincent

In the «blue» region of the Trp emission spectrum the model of two or three lifetime components was fitted to the experimental data (Table 9.1). In the "red" region only one exponential decay was discovered, but the model was poorly fitted; the model was corrected by including negative preexponential term or skipping the rising part of the decay curve.

Table 9.1. Distributions of fluorescence lifetimes of indol derivatives in glycerol, ethanol and aqueous solution

Fluoro-phore	Solvent	E_m [nm]	τ_1 [ns]	τ_2 [ns]	τ_3 [ns]	a_1	a_2	a_3	τ_{av} [ns]
Trp	Glycerol	320	0.4	1.5	5.2	0.39	0.18	0.43	2.4
		350	n	2.2	5.3	n	0.16	0.84	4.9
		390	n	n	5.3	n	n	1.00	5.3
Trp	Ethanol	320	n	0.4	2.5	n	0.30	0.70	1.9
		350	n	0.8	2.5	n	0.15	0.85	2.3
		390	n	n	2.5	n	0.16	0.84	2.4
Trp	Water	320	n	0.6	3.2	n	0.33	0.67	2.2
		350	n	0.6	3.1	n	0.15	0.85	2.8
		390	n	n	3.2	n	n	1.00	3.2
NATA	Glycerol	320	0.3	1.1	5.2	0.47	0.19	0.34	2.1
		350	n	1.5	5.3	n	0.15	0.85	4.8
		390	n	n	5.3	n	n	1.00	5.3
NATA	Ethanol	320	n	0.6	4.2	n	0.04	0.96	4.4
		350	n	0.3	4.2	n	0.03	0.97	4.0
		390	n	n	4.2	n	n	1.00	4.2
NATA	Water	320	n	0.5	3.1	n	0.06	0.94	2,9
		350	n	0.5	3.1	n	0.03	0.97	3.0
		390	n	n	3.1	n	n	1.00	3.1
NA-Trp	Glycerol	320	0.2	1.1	5.8	0.50	0.21	0.29	2.0
		350	0.8	1.7	5.9	0.17	0.04	0.79	4.9
		390	n	n	6.3	n	n	1.00	6.3
NA-Trp	Ethanol	320	n	1.1	5.1	n	0.07	0.93	4.8
		350	0.4	2.0	5.5	0.05	0.07	0.88	4.8
		390	n	n	5.1	n	n	1.00	5.1
NA-Trp	Water	320	0.9	n	4.7	0.15	n	0.85	4.1
		350	0.3	1.6	5.0	0.05	0.13	0.82	4.2
		390	0.2	1.7	4.9	0.07	0.06	0.87	4.2
Gly-Trp	Glycerol	320	0.4	1.7	5.3	0.66	0.12	0.22	1.6
		350	0.4	1.6	5.3	0.16	0.19	0.65	3.8
		390	n	n	5.5	n	n	1.00	5.5
Gly-Trp	Ethanol	320	0.8	2.0	4.1	0.50	0.37	0.13	1.6
		350	0.6	1.8	4.1	0.40	0.45	0.15	1.5
		390	n	1.0	3.2	n	0.62	0.38	1.9
Gly-Trp	Water	320	0.3	1.1	n	0.36	0.64	n	0.8
		350	0.2	1.2	n	0.38	0.62	n	0.8
		390	0.3	1.2	n	0.36	0.64	n	0.9

Table 9.1. (cont.)

Trp-Phe	Glycerol	320	0.4	1.6	4.4	0.43	0.19	0.38	2.2
		350	n	1.9	4.7	n	0.25	0.75	4.0
		390	n	n	4.7	n	n	1.00	4.7
Trp-Phe	Ethanol	320	0.3	2.1	6.3	0.48	0.30	0.22	1.1
		350	0.4	2.1	6.3	0.12	0.37	0.49	3.9
		390	n	2.0	6.3	n	0.34	0.66	4.8
Trp-Phe	Water	320	0.6	1.9	6.3	0.26	0.69	0.05	1.8
		350	0.4	2.0	6.6	0.31	0.66	0.03	1.6
		390	0.3	2.1	n	0.22	0.78	n	1.7
Indole	Ethanol	320	n	n	3.7	0	0	1.00	3.7
		350	n	n	3.7	0	0	1.00	3.7
		390	n	n	3.7	0	0	1.00	3.7

Excitation is 280 nm, slits are 10 nm; emission slits are 5 nm. 90% glycerol, 90% ethanol, and aqueous solution (a 10 mM phosphate buffer, pH 7.1) were used at 20°C. NATA is N-acetyl-tryptophan-amide, NA-Trp is N-acetyl-tryptophan, Trp is L-tryptophan, Gly-Trp is glycyl-tryptophan, Trp-Phe is tryptophyl-phenylalanine, E_m is the emission wavelength, $\tau 1$, $\tau 2$ and $\tau 3$ are the lifetime peaks, $a1$, $a2$, and $a3$ are their amplitudes, τ_{av} is the average lifetime. The data represented without a negative component. Decay measurements were carried out using the UV synchrotron radiation (Orsay, France), in collaboration with J.Gallay and M.Vincent.

The negative subnanosecond peak is caused by rapid formation the Trp-solvate excited-state complexes. When these complexes are formed, the energy is not emitted but stored. The initial part of the total decay curve contains a rising component, which belongs to the emission from accumulated excited-state complexes. The same process takes place also in polar solutions with other fluorophores. In «blue» region two or three lifetime components give a good fit without any additional negative component.

A negative component was also found in the fluorescence decay of some proteins [230–232, 530, 531]. In the "red" region of the emission spectrum, the fit of the data at the rising edge of the decay curve is improved by the use of negative preexponentials. The rapid formation of an excited-state complex, probably involving one or several water molecules in solvation shell, was assumed [230, 231].

The increase in the average lifetime with the emission wavelength is a result of the change in the amplitudes of the lifetime components (Table 9.1). The lifetimes which correspond to these components are constants and almost do not depend on wavelength (the minor variations are due to errors in the experiment and analysis). This fact cannot be described within the known "continuous" model of universal molecular interactions. The greater the number of different types of interactions, the greater the number of the lifetime components. Amplitudes of the lifetime components are in proportion to contributions from the different excited-state populations.

A rapid deactivation of an excited Trp by OH-groups is observed in the Trp-solvate complexes. This process leads to shorter lifetimes as compared with the lifetimes in apolar solvents or vacuum. Quenching fluorophores by polar solvent is a dynamic process, which depends on temperature and viscosity. The formation of emitting complexes with OH-groups successfully competes with deactivation by these groups.

In apolar organic solvents all fluorophore molecules emit in «blue» region as compared with emission in polar solvents. This «blue» emission is completely monomeric, since fluorophore-solvent interactions are negligible. Unlike this, in polar solvents a polar fluorophore molecule strongly interacts with its solvate shell. A ground-state interaction leads to some red shift in absorption on 100-1000 cm^{-1}. More strong interaction can arise after photoexcitation. A photoexcited fluorophore with a large dipole moment either transfers a part of energy to the solvate shell molecules or is deactivated by they. In the first case the emission band shifts to «red» region, up to ~4000 cm^{-1}. The «red» structureless emission band belongs to the fluorophore-solvent excited complexes. A small «blue» component of emission in polar solvents could be formed by a minor population of fluorophore molecules which are slightly interact with the solvent shell.

The emission of apolar fluorophores does not shift with polarity of a solvent, since the dipole moment of apolar fluorophore and interaction with a solvent are negligible. For instance, the position of fluorescence band of phenol depends on the polarity of a solvent, but that of benzene does not.

The red shift in the Trp emission is correlates with the solvent dielectric properties and the refractive index. However, this correlation does not hold for hydrogen-bonding solvents [233]. The red shifts of the 3-methylindole fluorescence in polar solvents (water, methanol, butanol) were calculated in [234], using a hybrid method that couples molecular dynamics and a semiempirical molecular orbital procedure. Recently J.Vivian and P.Callis predicted the emission wavelengths of 16 proteins using a hybrid quantum mechanical-classical molecular dynamics method with the assumption that only electrostatic interactions affect the transition energy [532].

In paper by D.Toptygin et al. [530] a criterion for distinguishing between homogeneous and heterogeneous time-resolved fluorescence emission of tryptophan residues in proteins was suggested. The width at half-maximum of the instantaneous emission spectra is a constant for a homogeneous system and is not a constant for a heterogeneous system. Also, microscopic dielectric response function in two proteins where found. It is important to stress here that the spectral heterogeneity of emission of single-tryptophan proteins can be caused by a) conformers, existed initially and b) photoconformers, appeared due to photoexcitation.

9.2 TRP and NATA in Water

For 3-methylindole in water the solvent relaxation after excitation appears to create a red shift of 3800 cm^{-1} [235]. However, this shift in emission considerably differs from the simulated values calculated using dipole moments in frameworks of a hybrid method consisting of a semiempirical quantum mechanics with singly excited configurations for the fluorophore and classical molecular dynamics for the solvent [235].

The red shift in the Trp emission in water (maximum at ~350 nm) as compared with emission in apolar solvents (maximum at ~310 nm) is equals to ~3600 cm^{-1}. This value coincides with the frequency of valency OH-vibrations of H_2O, i.e. with the most intense and wide band in the IR absorption spectrum of water [236, 237]. According to [238, 239] it could be assumed that the red shift depends on the intramolecular vibrations of the solvate shell accepting the vibrational energy from the photoexcited Trp molecule. This accepting is result of strong interactions between an excited Trp and its water shell.

A photoexcited Trp molecule either gives to the water shell a part of energy or all energy. In the first case (fractional energy transfer, Chapter 10) the emission band shifts to «red» region. In the second case (deactivation) the emission disappears. Using jet cooled fluorescence, a sharp fall in the hydroxyindole lifetime as compared with indole was shown [240]. The fall depends on the position of hydroxy-group: 17 ns in indole, 11.1 ns in 5-hydroxyindole, and ~0.2 ns in 4-hydroxyindole. In the Trp aqueous solution the both ways, fractional transfer and deactivation are realized. The formation of red-emitting complexes with OH-groups successfully competes with deactivation by these groups. It leads to a red shift of emission and shortening in lifetime as compared with apolar solutions.

Tryptophan is anisotropic zwitter-ion consisted of apolar indole chromophore and polar chain. Therefore, at least two types of interactions of this molecule with solvent are possible. In aqueous solution Trp displays spectral heterogeneity of lifetimes (Table 9.1). In the "blue" region of the emission spectrum Trp exhibits two excited-state populations: a dominant one has a long lifetime (3.2 ns) and a minor - a short lifetime (0.6 ns). These lifetime values almost coincide with previous data by Szabo and Rayner [223]. Using decay measurements, these authors resoluted the steady-state emission spectrum of tryptophan into the two bands: the major 3.1-ns long-lifetime component with maximum at ~360 nm, and the minor 0.5-ns short-lifetime component with maximum at ~335 nm. The proportion between the both components depends on the temperature and pH [241]. As is seen from Table 9.1, only long-lifetime component is observed in the «red» region. Here the fit was not enough appropriate. Better results could be obtained using maximum entropy method with subnanosecond negative preexponential terms (see below). The long-lived 360-nm band is formed due to intramolecular valency vibrations of the OH-group in water molecules of the solvate shell accepting the vibrational energy of ~3600 cm^{-1} from the photoexcited Trp. A small «blue» lifetime component is formed by a minor population of photoexcited Trp molecules which interacts with other IR frequencies of water (probably 1700 and 900 cm^{-1}). As was shown in [463], the both decays increase by a factor of two in deuterium water. At least 83% of the enhancement is due to the exchange of the ammonium hydrogens by deuterium [233]. Therefore, the ammonium group acts as intramolecular quencher.

The Trp decay times and their preexponential terms are constant as a function of excitation wavelength (from 250 to 295 nm) [241]. This indicates that the absorption spectra, which could be associated with the two hypothetical ground-state populations, overlap completely and the contribution of both populations to the total absorption remains constant over the whole absorption spectrum [241]. Therefore, the two populations appear after photoexcitation. As is known, photoexcitation sharply increases the dipole moment of Trp. This leads to more strong interactions in the Trp-$(H_2O)_n$ excited-state complexes than that in the ground state.

NATA also has two components in "blue" region. The amplitude of short lifetime component in the NATA fluorescence decay is less than that in the decay of tryptophan: 6 % and 33 %, respectively, at the emission wavelength of 320 nm. In previous studies [223, 225] the minor lifetime component of NATA was not detected. As is seen from Table 9.1, only single long-lifetime component is observed in the «red» region.

9.3 TRP, NATA and Indole in Ethanol

Two excited-state populations were observed in the emission from Trp in ethanol in the "blue" spectral region (Table 9.1). The appearance of two emission components is result of two types of the excited-state interactions with ethanol. One long-lifetime excited-state population and a fast (0.04 ns) component of a negative amplitude were observed in «red» region. For NATA the results are similar except that at 320 nm the amplitude of the fast decay component of NATA is a lower than that of Trp. The average value of the long-lifetime population is greater for NATA than for Trp (4.2 and 2.5 ns, respectively).

Indole is a simple slightly-polar molecule with small dipole moment. Unlike Trp or NATA, it has not a polar fragment. Therefore, only one type of interaction between indole and ethanol is possible. The indole emission should be homogeneous. In fact, the lifetime of indole in ethanol was not changed along the emission spectrum in the wavelength range between 320 and 390 nm (the decay of indole in water and in glycerol was not measured since indole is aggregated in these solvents). A very narrow peak at 3.7 ns was revealed in the "blue" as well as in the "red" region. The poor fit was corrected [219, 228] using negative preexponential terms or skipping the rising part of the decay curve. In course indole fluorescence in ethanol at 310 nm the main lifetime component of 3.7 ns and a minor very short-lived component of ~0.04 ns were determined in [266, 500]. However, the picosecond positive and negative components belong to the «noncorrect» time interval outside a good time resolution of the used technique (a full-width of the light pulse at half maximum was ~0.5 ns at the 8.33-MHz frequency [228]). Therefore, these components could be an artifact induced by: (a) noncorrect subtraction of an excitation pulse, (b) a scattered light, (c) small signal/noise ratio, and (d) poor analysis.

9.4 TRP and NATA in Glycerol

Three excited-state populations are observed in the "blue" region of the emission spectrum from Trp or NATA in glycerol (Fig. 9.2 and Table 9.1). The lifetime values of both molecules are similar. The first lifetime population is centered around the values of 0.4 or 0.3 ns and the second one is located within 1.5 or 1.1 ns. The third lifetime population is observed at 5.2 ns.

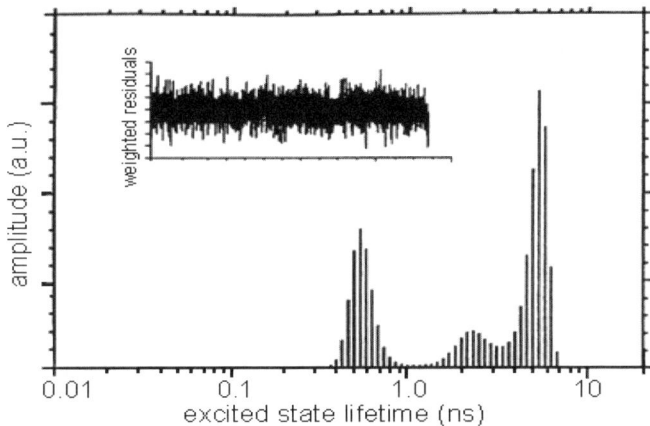

Fig. 9.2. Fluorescence lifetime distribution of tryptophan in 90% glycerol at 330 nm [219]. The excitation wavelenght is 280 nm. Fluorescence intensity decays were measured using time-correlated single photon counting technique. Decay as a sum of 150 exponentials was analyzed by the method of maximum entropy/ The inscrt shows the weighed residuals

Only two populations, the second and the third, are observed in the maximum of emission spectrum; the first population disappears (Table 9.1).

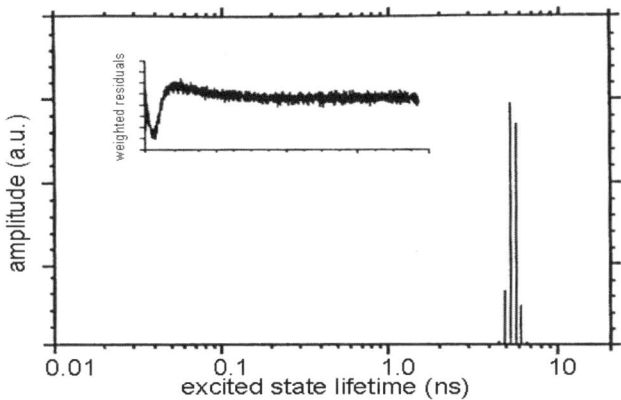

Fig. 9.3. Lifetime distribution of fluorescence of tryptophan in 90% glycerol in the "red" region (at 385 nm) using only positive preexponential terms [219]

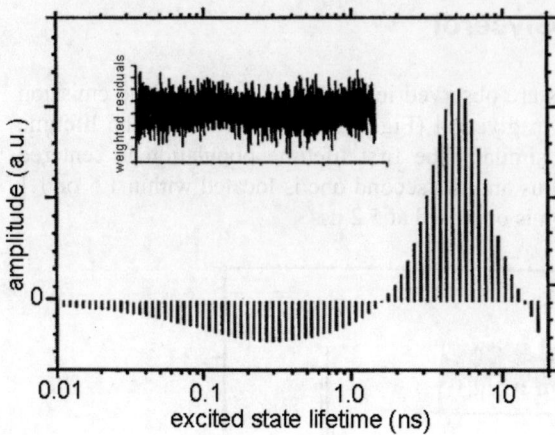

Fig. 9.4. Lifetime distribution of fluorescence of tryptophan in 90% glycerol in the "red" region (at 385 nm) when negative amplitudes were used [219]

In the "red" region of the emission spectrum only the long lifetime excited state population is observed. However, in this spectral region the data are not rationalized by a sum of exponentials with positive amplitudes (Fig. 9.3). An acceptable fit is obtained when negative amplitudes are used (Fig. 9.4) or when the rising part of the experimental decay curve is skipped (Fig. 9.5).

The average lifetime of Trp in glycerol increases from 2.4 ns in the "blue" region to 5.3 ns in the "red" region (Table 9.1). This result coheres with the one obtained in the phase-modulation measurements [167] on the same system (Chapter 8). The average lifetime of NATA increases in a similar way. The increase in the average lifetime with the emission wavelength is a result of the change in the amplitudes of the components.

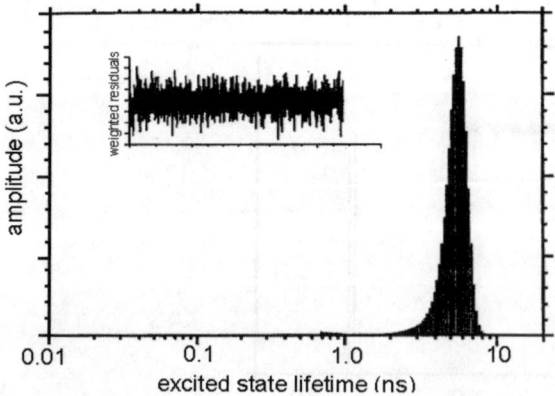

Fig. 9.5. Lifetime distribution of fluorescence of tryptophan in 90% glycerol in the "red" region (at 385 nm) using only positive preexponential terms, the fit starts 1.5 ns after the peak of the excitation flash [219]

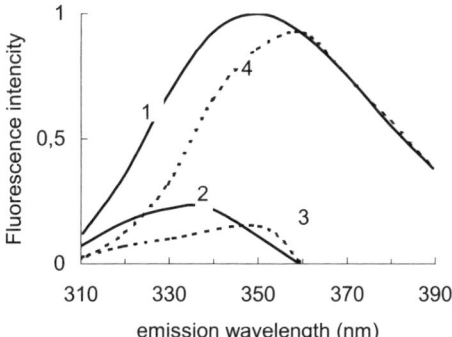

Fig. 9.6. Resolution of the steady-state emission spectrum of NATA in 90% glycerol (1) into spectral components corresponding to different lifetimes: the shortest lifetime component (2), the intermediate one (3) and the long-lived one (4) [219]

There is no evidence of a steady-state spectral heterogeneity of Trp or NATA in glycerol: excitation spectra recorded while monitoring the emission at 320 nm and 390 nm are superimposable, and the emission spectra obtained using the excitation wavelengths of 280 and 295 nm coincide (see Figures in Chapter 8).

Fig. 9.6 shows decomposition of the steady-state emission spectrum of NATA in decay-associated spectral components. Two minor components and a major component are overlapped in «blue» region. Similar results of decomposition were obtained for Trp. The three lifetime components are result of three types of excited-state interactions with the solvent. Probably, three OH-groups in the glycerol molecule (and also OH-groups of the water admixture, 10 %) could accept three different portion of a vibrational energy from the fluorophore.

The lifetime distributions obtained for Trp and NATA in glycerol at 4°C were similar to those obtained at 20°C [219]. Increasing the temperature to 50°C led to an increase in the amplitude of the short-lived component and to a decrease or even to disappearance of the intermediate lifetime component. The long-lived component decreases from 6 ns at 4°C to 3.2-3.4 ns at 50°C. The lifetime distribution in glycerol at high temperature was similar to that obtained in ethanol or water at room temperature (not shown). The increase in the temperature decreases the viscosity that leads to more rapid deactivation: the long-lived lifetime is diminished.

9.5 Dipeptides

Interactions in dipeptides are represented not only by fluorophore-shell interactions, but also intramolecular interactions. This results in complex lifetime distributions. The average three components obtained for dipeptides from 3 to 5 experimental components are shown in Table 9.1. The lifetimes of dipeptides in water or ethanol and their amplitudes differ from those of Trp or NATA. The distributions for dipeptides in glycerol are simpler than distributions in water or ethanol. Therefore, interactions between the Trp residue and the second amino acid residue of the dipeptide are viscosity-limited. In viscous media (glycerol) these interactions are weak. Trp residue actively interacts with a second amino acid of the dipeptide when a flexible photoconformer is formed. The viscosity prevents the formation of photoconformers.

9.6 Exciplexes in Proteins

Indole, an analogue of Trp, can form excited complexes with alcohols (and with other polar organic molecules) [57, 227, 242]. These excited complexes were named as «exciplexes» by analogy to known exciplexes of aromatic hydrocarbons.
The word «exciplex» strictly applied to complexes, which arise due to photoexcitation (without any stable complex in the ground state) and consist of one vibrational energy acceptor per one photoexcited donor. In n-hexane or other apolar organic solvent the formation of the indole-alcohol exciplexes at collision of photoexcited indole with alcohol molecule is accompanied by quenching the monomer indole fluorescence at ~310 nm and appearance of a structureless broad band of the exciplexes in more "red" region. For example, the emission maximum of 3-methylindole in n-heptan or methylcyclohexane lies at 308 nm; the spectrum is narrow and has vibrational structure [243]; adding small portions of butanol leads to decrease in the intensity of the 308-nm band and appearance the structureless exciplex band at 316 nm (one butanol molecule per one methylindole molecule in the complex) and ~335 nm (the 2:1 complex) [188, 227, 242]. This picture is similar with those for Trp in polar solvents (water, ethanol, or glycerol). Moreover, addition of small portions of butanol results in an apparent broadening of the absorption spectrum with an increased absorbance at >290 nm [67, 227]. This indicates a complex formation in the ground state. Therefore, the excited complexes are not «classical» exciplexes.

The fluorescence of proteins and nucleic acids is mainly of exciplex character [227, 242, 244, 245]. A photoexcited chromophore forms an unstable electronically excited complex, the exciplex, with neighbouring chromophores or groups. Exciplexes may be fluorescent or non-fluorescent [53, 47]. Therefore, if new band or the "red" shift are absent when the fluorescence of tryptophan or tyrosine is quenched by the intrinsic groups in a protein or by external quenchers, it does not prove the absence of an exciplex. The formation of emitting exciplex is the primary reason for the red-shift and broadening of the fluorescence spectrum of proteins. The emission band of the Trp exciplex is shifted towards longer wave lengths and broadened. The shift distance between the "blue" and "red" components usually is ~1000–3600 cm^{-1}. These values are close to the frequencies of the IR vibronic transitions. The emission of these exciplexes could be described by the model of fractional energy transfer [238, 239]: vibronic modes of the ground state of the «quencher» accept some vibrational energy from the photoexcited chromophore (IR-transitions in the range of 3600–1000 cm^{-1}) and the rest energy is emitted in the "red" region (Chapter 10).

Fig. 9.7 demonstrates the emission spectra of endonuclease at different temperatures [242]. At low temperature (150 K) the mobility of the amino acid residues is almost zero because the protein is frozen. Thus, the emission band of tryptophan is "blue" positioned and structured. This emission is of the monomer nature. The mobility increases with an increase in temperature. Hence, the monomer fluorescence is quenched, the spectrum shifts towards the "red" region and loses the vibrational structure. The appeared emission is of the exciplex nature. Three types of the Trp exciplexes are formed: (a) exciplexes with polar amino acid residues, (b) exciplexes with a bound water, (c) exciplexes with propylene glycol and ethylene glycol. Similar spectra were obtained for NATA in a mixture of propylene glycol, ethylene glycol and water under the same conditions [242].

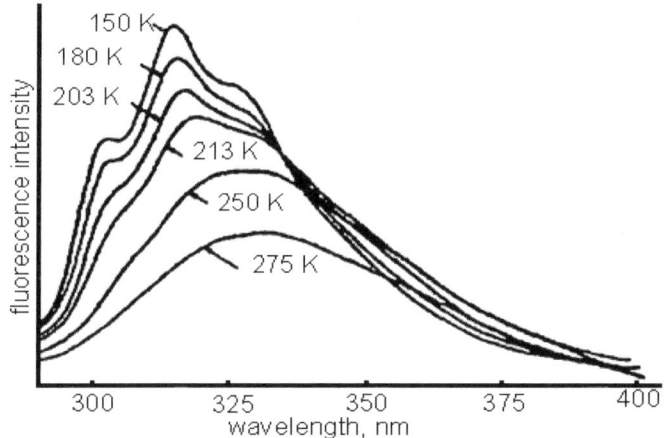

Fig. 9.7. Fluorescence of endonuclease (isolated from *Staphylococcus aureus*) in a mixture of propylene glycol, ethylene glycol and water (1:1:2) with a 10 mM phosphate and 0.1 M NaCl (pH = 7) at different temperatures [242]

One of the first time-resolved studies of single-tryptophan proteins using nonlinear least-squares regression analysis and the maximum entropy method was made with horse heart apocytochrome-c [229]. Four separate lifetime components were discovered: 0.2, 1, 3 and 5 ns. As a function of the emission wavelength, these values remained constant, whereas the amplitudes were changed. The contribution of the short-lived component increased in the «blue» region of the emission spectrum. Apocytochrome-c is not native protein. The authors suggested that the protein exist under different conformational states with different lifetimes [229]. Five lifetime components in a single-tryptophan synthetic polypeptide and porine inserted into detergent micelles or lipid membranes were detected in [246]. One of them was a minor long-lived 20-ns component with amplitude of ~0.002.

Fluorescence lifetimes of three native single-tryptophan proteins of known structures are given in Table 9.2.

Table 9.2. Lifetime distributions of fluorescence decay for ribonuclease T1, parvalbumin-Ca^{2+} and porcine phospholipase A2

Protein	Em. [nm]	$t1$ [ns]	$t2$ [ns]	$t3$ [ns]	$a1$	$a2$	$a3$	t_{av} [ns]
Ribonuclease								
	310	n	n	4.0	n	n	1.00	4.0
	320	n	n	4.0	n	n	1.00	4.0
	340	n	n	4.0	n	n	1.00	4.0
	375	n	n	4.0	n	n	1.00	4.0
Parvalbumin								
	310	n	1.1	3.4	n	0.15	0.85	2.7
	320	n	1.3	3.3	n	0.07	0.93	3.3
	340	n	1.7	3.3	n	0.05	0.95	3.3
	375	n	n	3.4	n	n	1.00	3.4
Phospholipase								
	320	0.6	2.1	5.3	0.71	0.24	0.05	1.2
	350	0.7	2.7	7.2	0.63	0.32	0.05	1.7
	385	0.7	2.7	6.5	0.61	0.33	0.06	1.7
Phospholipase in 90%glycerol								
	320	0.4	2.1	5.6	0.40	0.43	0.17	2.0
	350	0.8	2.7	6.3	0.19	0.54	0.27	3.3
	385	n	n	4.0	n	n	1.00	4.0

Excitation is at 295 nm, slits are 5 nm. Solutions used at 20°C: a 100 mM acetic buffer (pH 5.5) for ribonuclease, a 25 mM Tris-HCl buffer with 1 mM calcium chloride (pH 8.2) for parvalbumin, a 100 mM acetic buffer (pH 5.8) for phospholipase. Multipass cuvettes were used at "blue" emission wavelengths. The method of correlated spectroscopy was used (in collaboration with J.Gallay and M.Vincent).

The decay of RNAase T1 is single-exponential and does not depend on the emission wavelength. This is in accord with the single-exponential decay curve obtained earlier by many authors. The tryptophan residue of this protein is buried in the globula and rigidly fixed. The tryptophan residue and its microenvironment are not flexible, and only a single conformational state exists. Therefore, the tryptophan residue forms only one emission center.

The decay of cod parvalbumin is double-exponential in the "blue" region and depends on the emission wavelength. No short lifetime component is observed in the "red" region. In parvalbumin-Ca^{2+} the minor short-lifetime "blue" component corresponds to the monomeric emission, and the long-lifetime "red" component corresponds to the emission of the exciplex. From the data of Table 9.1 and 9.2 and Fig. 9.8 (see parvalbumin lifetime distributions in Figures in Chapter 6) we may conclude that at room temperatures the tryptophan residue of parvalbumin forms an exciplex with any polar amino acid residue. Probably, this polar amino acid residue can moved to the photoexcited tryptophan and interacts with it.

No interaction exists in the ground state, since the absorption spectrum of parvalbumin has fine vibronic bands (Fig. 9.9) similar to the absorption of free tryptophan in a vacuum. Therefore, the interaction appears after photoexcitation, when a protein photoconformer is formed.

9.6. Exciplexes in Proteins 87

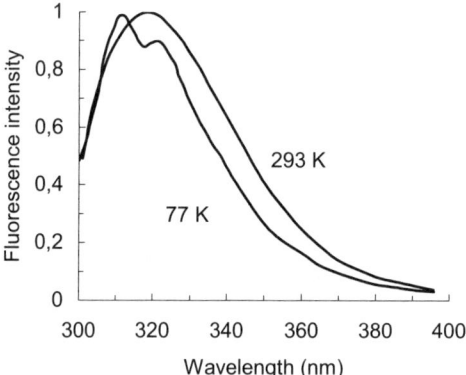

Fig. 9.8. Fluorescence spectra of calcium-loaded parvalbumin of cod in 20 mM HEPES + citric acid buffer with pH 6.8 at 293K (excitation at ~297 nm) and 77K (excitation at ~290 nm). Data were obtained by G.Deikus

The fluorescence decay of phospholipase A2 in aqueous solution has three life time components of different amplitudes. Probably, several different quenching groups are located near the single tryptophan residue. Since the tryptophan residue in phospholipase is outside the globule [119], water may serve as one of the quenchers. In fact, the short and the middle lifetimes (0.7 and 2.7 ns) are close to the two lifetimes of Trp or NATA in water. For phospholipase in glycerol the lifetimes are very similar to those in water but the amplitude of the long lifetime component is larger. The short- and the middle-lifetime components disappear in the "red" region (at 385 nm). In bviscous media (glycerol) the conformational mo-

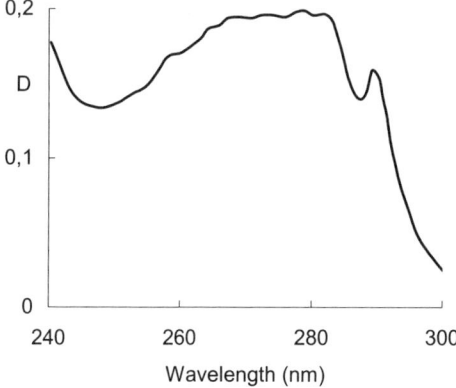

Fig. 9.9. Absorption spectrum of calcium-loaded parvalbumin of cod at 293K (own unpublished data)

bility of phospholipase vanishes. This leads to a decrease in the quenching by OH-groups of water, glycerol, and by polar groups of the protein.

Only in several proteins of a very rigid structure and a non-polar microenvironment near tryptophan residues the emission is of monomer type with "superblue" maximum at 308–315 nm. Spectra of these proteins are narrow and have vibronic structure, for instance, in apoazurin [243].

Two mutants of the copper protein azurin were prepared in [247]. The mutation was carried out to increase the mobility and change emission properties of Trp-48. In the wild-type protein Trp-48 is located in a highly hydrophobic and rigid environment. The emission of Trp-48 in the wild-type protein has a fine structure and is strongly quenched by copper. The tryptophan fluorescence in mutants is drastically influenced by the introduction of a polar residue close to tryptophan. The emission becomes red-shifted, structureless and dependent on the excitation wavelength. Previously some authors explained this effect with azurin as a result of electron transfer from the excited state of tryptophan to copper. Fluorescence lifetimes of substituted indoles in solution and in free jets were measured in model experiments [226]. The authors got some evidence for the charge-transfer mechanism of intramolecular quenching. However, unlike the electron-transfer hypothesis, resonance Raman spectroscopy data do not indicate the formation of a tryptophan radical cation intermediate in azurin [247].

In proteins the exciplexes are formed not only by tryptophan residues. For instance, with use of time-resolved polarized fluorescence spectroscopy the exciplex is shown to be formed by flavin and Tyr in the lipoamide dehydrogenase and glutathione reductase [248].

9.7 Conclusion

Tryptophan, its derivatives and dipeptides in polar solutions, as well as some proteins, show discrete fluorescence heterogeneity along emission wavelengths. Two or three lifetime components are observed in the "blue" emission region. In the "red" region the number of lifetime components is usually smaller than in the "blue" region (often a single long-lifetime component is found), but the model is poorly fitted. The initial part of the total decay curve contains a rising component, a subnanosecond peak of the negative amplitude, which belongs to the emission from accumulated excited-state complexes (exciplexes).

The discrete emission heterogeneity cannot be explained completely within the traditional models of «continuum» relaxation and ground-state conformers. Discrete relaxations and flexible photoconformers should be taken into account. The amplitudes of lifetime components are in proportion to contributions from the different excited-state populations. The greater is the number of interactions of different types, the greater is the number of lifetime components.

A rapid dynamic deactivation of an excited Trp by OH-groups or amino acid residues leads to a shortening in lifetime. The formation of red-emitting complexes with OH-groups or amino acid residues successfully competes with the deactivation. It leads to a red shift and broadening of the emission spectrum. The red shift depends on the intramolecular vibrations in the solvate shell or closest amino acid residue accepting the vibrational energy from the photoexcited Trp. This accepting is result of strong interactions in exciplexes.

10 Mechanisms of Exciplex Formation

The physical mechanism of the energy conversion in exciplexes in biopolymers is not yet completely clear because the red shift in the exciplex emission relative to the monomer is rather ambiguously explained even for simple model systems [57, 162, 249].

10.1 Generally Accepted Models

The viewpoints about the mechanisms of energy transformation in exciplexes were proposed in the 60's and 70's, when the classical mechanism of energy transfer was recognized. These mechanisms assumed the existence of a resonance, based on spectral overlap. To explain the phenomena occurred in the course of the exciplex formation, when the electronic spectra of the donor and acceptor did not overlap, new mechanisms were suggested (without resonant energy transfer) [250]. The currently accepted mechanism is as follows [52–55, 57]. An electron is transferred to a free orbital of the oxidizer after the excitation of the first molecule («reducer») and its subsequent collision with the second one («oxidizer»). Then from this orbital the electron is transferred to the ground energy level of the reducer, which is accompanied by emission of a photon. Such a "reversible" electron transfer was explained in terms of redox and ionization potentials. Experimental data on singlet and triplet excimers and exciplexes are reviewed in [57, 249].

For instance, intramolecular electron transfer in the excited state from a permethylated hexasilane chain to pyrene was observed [251]. In addition to the emission from the locally excited state, the fluorescence spectrum showed a solvent-dependent exciplex band. Fluorescence from the pairs anthracene–dimethylaniline and perylene-dimethylaniline was studied in supersonic jets [252]. A charge-transfer exciplex emission was assumed to be formed by intramolecular energy transfer within the Van der Waals complex from a locally excited state of the hydrocarbon acceptor. The exciplex emission bands (measured 150 and 25 ns after the laser pulse) were broad and structureless; the maximum was located at 455 nm for the anthracene–dimethylaniline pair and at about 460 nm for the perylene–dimethylaniline pair. It was suggested [252] that in the jet-cooled system emission was occurred probably due to vibrationally excited exciplexes, while in liquid solution it was most likely from vibrationally relaxed ones. The photophysical and photochemical behaviour of (N,N-dimethylamino)styrenes were studied in [253]. Unlike to exciplexes between methylstyrene and triethylamine, the decay time of their intramolecular exciplex increased with the solvent polarity.

Exciplex formation leads to the quenching of the luminescence from the excited molecule and to the appearance of a red-shifted wide structureless luminescence band, which does not belong to the quencher [52–55]. This structureless band belongs to the exciplex as a whole. In solutions the quenching is also accompanied by a decrease in the lifetime of the monomer and by changes in the decay kinetics. The shape of the monomer luminescence spectrum remains unchanged. In the excitation spectrum of the exciplex the monomer band is clearly visible. The intensity of exciplex luminescence is less than or equal to the intensity loss in the luminescence of the quenched monomer. The exciplex luminescence in solutions is highly depolarized. The quenching occurs when the monomer is close to the quencher, and the effect sharply depends on the distance. Thus, we may conclude that an energy transfer takes place here (Chapter 11). Unlike the classical resonance energy transfer, this is non-resonance fractional energy transfer.

The exciplexes of indole with alcohols are well known [57, 227, 242]. The formation of tryptophan-solvent exciplexes was experimentally proved by many methods, and recently by time-resolved photoacoustic spectroscopy [254]. It was emphasized that «in exciplexes of indoles with polar molecules, the exciton interaction is hardly possible; the charge-transfer interaction cannot be significant as well » [57]. In fact, the conventional models of excitonic resonance interactions [52, 57] and reversible electron transfer [52–55, 57] are not appropriate here. Neither resonance between electronic levels nor effective electron-transfer is possible. These models fail to predict the value of the red shift of the structureless excimer or exciplex emission relative to the monomer fluorescence band, and sometimes these models contradict the experimental results even at the qualitative level.

At present time, the red shift of emission and the form of bands in polar solvents are calculated by a quantum mechanical and classical molecular dynamics methods [235, 255]. For instance, solute-solvent interaction and dynamics of indole and 3-methylindole in water at 300 K were simulated in [235]. About 50% of the Stokes shift occurs in ~15 fs with a Gaussian response function, and the another exponentially decays is with the time =400 fs. The authors concluded: "There is no need to invoke a specific exciplex formation involving charge transfer from the indole to water molecules to account for the large Stokes shift in water. The large fluorescence shift can be completely explained by dipole-dipole interactions at distances beyond the Van der Waals radii".

The non-specific interactions between the solvent and chromophore molecules affect the energy difference between the ground and the excited states. The frequency shift between absorption and emission is expressed by the Bakhshiev's equation [82]:

$$\Delta v = 2h^{-2}c^{-2}a^{-3}(\mu^2 - \mu^{*2} - 2\mu\mu^*\cos\alpha) \times$$
$$\times [(\varepsilon-1)/(\varepsilon+2) - (n^2-1)/(n^2+2)](2n^2+1)^2/(n^2+2)^2 + \text{const} \qquad (10.1)$$

where Δv is expressed in cm^{-1}, n is the refractive index of solvent, ε is the dielectric constant, μ and μ^* are the dipole moments of the chromophore molecule in the ground and excited states, a is the cavity radius for the chromophore molecule, h is the Planck constant, c is the velocity of light, α is the angle between μ and μ^*. The Bakhshiev model describes the red-shift of emission in polar solvents in terms of continuous interactions with solvent shell. This model is more good approximation than the known Lippert model [67]. As was mentioned in [67], the both models are phenomenological ones.

Refractive index depends on the wavelengths along the UV, visible and IR scale. The largest changes in n take place at maxim of the absorption bands. In the Bakhshiev model, n is an equivalent of the solvent absorption transitions. As known, the complex refractive index may be separated in two components: the natural component, and imaginary one. The last corresponds to the light absorption [82].

10.2 Fractional Energy Transfer in Exciplexes

It was found out long ago that quenching of electronically excited molecules by various compounds can be accompanied by a substantial population of vibrational levels of the quencher molecule [256], and that the fast internal conversion in polyatomic molecules is a result of the deactivating effect of their own vibrational modes, in particular, the valence vibrations of the OH- and CH-groups [257–259].

Behaviour of the luminescence spectra of exciplexes and excimers corresponds to the process of fractional energy transfer (Chapter 11). If the acceptor has an absorption band, which does not overlap with the donor luminescence, but is located at longer wavelengths, then it is possible that a part of the energy from the excited donor can be transferred, the rest energy is wasted as heat or emitted. For simple inorganic ions of rare-earth elements with discrete energy levels, the non-resonance transfer and emission of energy balance are observed [260]. "The donor may donate not all its excitation energy. If the donor has intermediate levels, it could relax on them and transfer the difference energy to the acceptor. The emission of photons of the difference frequencies is possible" [260].

The mechanism of the conversion of excitation in exciplexes and excimers could be divided into the following stages: (a) photoexcitation of the donor molecule; (b) vibronical relaxation in this molecule; (c) diffusional collision of donor and acceptor molecules during the lifetime of the donor; (d) the appearance of an electronic-vibrational interaction between the lower level of the electronically excited state (S1) of the donor and the vibrational levels of the ground state of the energy acceptor (these interactions lead to the appearance in the donor molecules of new levels, which lie 3000–4500 cm^{-1} below S1; these new levels, induced by acceptor, are called as the «intermediate» ones); (e) some portion of energy is transferred to the vibronic levels of the ground state of the acceptor (this process is accompanied by the relaxation of the donor to the intermediate levels); (f) the emission of photons of the difference frequencies which correspond to decay from the intermediate levels forms a structureless excimer or exciplex band; (g) the tra-

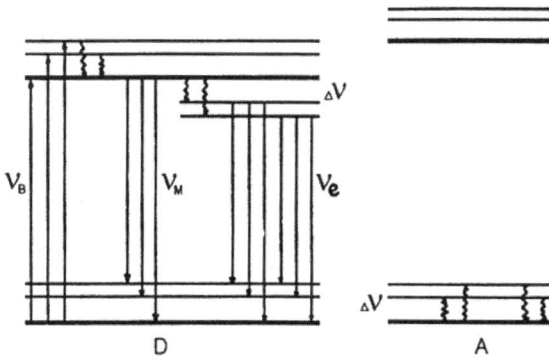

Fig. 10.1. Fractional energy transfer in exciplexes or excimers. D is the donor of energy (the monomer), A is the acceptor of vibrational excitation, v_b is the frequency of the excitation light, v_m is the frequency of the monomer emission, v_e is the frequency of the excimer or exciplex emission from the intermediate levels, Δv is the frequency of the vibrational modes that induce the "red" shift [238]

nsfer of the vibrational energy from the acceptor molecule to the solvent or the emission of IR quanta from the acceptor takes place.

Fig. 10.1 shows the processes occurred during the excimer or exciplex formation under the proposed fractional energy transfer [238]. According to this scheme, the acceptor lowers the S1 level of the donor. This process is similar to the lowering of the excited fluorophor levels by polar molecules of the solvent shell.

The scheme explains the broadening and structureless of excimer or exciplex emission band: several intermediate levels of donor molecules (Fig. 10.1) are responsible for the emission. The magnitude of the red shift, Δv, is mainly determined by the active vibrational transitions of the ground state of the quencher (vibronic acceptor).

The number of possible transitions (N) from intermediate levels in exciplexes can be obtained from the following equation:

$$N = WU \quad (10.2)$$

where W is the number of intermediate levels of the donor, which is determined by the number of vibronic levels of the ground state of the acceptor; and U is the number of vibronic levels of the ground state of the donor. For excimers $U=W$, thus:

$$N = U^2 \quad (10.3)$$

For the monomer fluorescence of the donor molecule $N=U$. Therefore, the excimer or exciplex band is structureless, broader and shifted forward the red region relative to the monomer fluorescence. The magnitude of the red shift depends on the most effective mean frequencies in the IR region of the acceptor molecules.

Multiatomic organic molecules have many absorption bands in the IR region and several electronic bands in the UV or visible regions. All these bands or some of them or only one band could interact with the donor. In the IR region, the most intensive bands (the O-H and C-H valency vibrations) are more active. Conversion of electronic excitation energy to vibrational modes in polyatomic molecules, in particular, in aromatic hydrocarbons, is verified by replacing hydrogen with deuterium [257], since the frequency and intensity of vibrations decreases after this replacement [126, 132]. Using jet cooled fluorescence, a sharp fall in the 4-hydroxyindole lifetime (~0.2 ns) as compared with indole (17 ns) was shown [240]; deuteration of 4-hydroxymdole to form -OD significantly lengthened lifetime. In deuterated solvents the fluorescence quantum yield and lifetime of fluorophores usually increase. For instance, when ethanol is replaced by C_2D_5OD, quantum yield of indole increase by >15% [238]; when H_2O is replaced by D_2O, lifetime of tryptophan increase by 2-fold [241]. The CH-vibrations in a solvent also quench the excited states. For instance, the lifetime of indole fluorescence in degassed cyclohexane solution is approximately two-fold low than in supersonic gas expansion [226].

The lowering in the energy levels of a donor when it donates vibrational quanta to an acceptor is similar to the lowering in the energy levels of electronically excited molecules when they interact with a polar solvent [82, 85]. It is well-known that polar solvents quench fluorescence. Moreover, the spectrum loses the fine structure, becomes broad, and shifts to the red region as compared to the spectrum in nonpolar solvents [67, 82, 138]. The same features are typical for exciplexes and excimers.

The proposed model of "fractional" energy transfer in exciplexes provides the explanation for the red shift, the appearance of the broad and structureless fluorescence band of tryptophan in polar solutions and proteins, as compared to the narrow and structured band of indole in hexane or azurine. The amino acid residues located near tryptophan are able to accept vibrational quanta from the electronically excited indole chromophore.

10.3 Exciplex of Aromatic Hydrocarbons

As is known, unlike the majority of excimers, exciplexes have large dipole moments. In polar solutions where the solvent shifts the spectra and quenches the luminescence, the model could not be tested. However, this model could be verified in non-polar solvents.

The fluorescence spectra of a number of aromatic hydrocarbons (pyrene, anthracene, dimethylanthracene, chrysene, diphenyloxazole, benzpyrene, etc.) and their exciplexes with diethylaniline were measured in n-hexane under aerobic conditions [238]. Hexane was used to minimize the effect of the solvent on the spectra. Oxygen eliminates phosphorescence of triplets. The emission intensity of some exciplexes is very low. The spectra were recorded in the «accumulation» regime of «SLM-4800»; the monochromator step was 0.5 or 1 nm; the spectra were smoothed and plotted on the necessary. Exciplex spectra were obtained by subtracting the initial fluorescence from the final one. The «red» shift, $\Delta v = v_m - v_e$ was

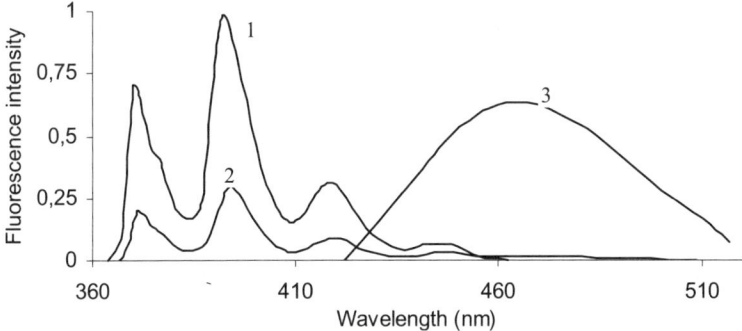

Fig. 10.2. Anthracene fluorescence (1); monomer and exciplex fluorescence after adding 30 mM diethylaniline (2); the exciplex fluorescence in a 33-fold enhanced scale (3) [130]. Excitation at 353 nm

calculated as the difference of frequencies between the "gravity" centers of the monomer (v_m), and the exciplex (v_e) bands. The magnitude of the red shift of the structureless exciplex band relative to the monomer band was about 3000 cm^{-1}.

For the indole-ethanol exciplex the red shift was about 3600 cm^{-1}; this value is close to the frequency of valency vibrations of the OH-group in alcohols, i. e. to the most intense and wide band of IR absorption. It was assumed [238] that the shift depends on the intramolecular vibrations of the quencher accepting the vibrational energy from the electronically excited donor. It was proposed that exciplexes are capable of emitting due to the radiative donor transitions from many intermediate levels.

Fig 10.2 shows the quenching of anthracene fluorescence by diethylaniline. The

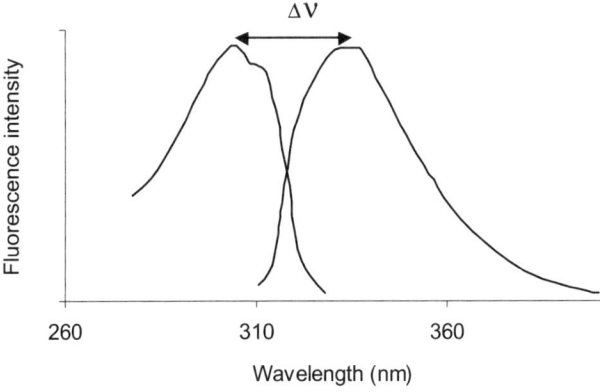

Fig.10.3 Corrected excitation (1) and emission (2) spectra of diethylaniline in *n*-hexane. Slits are 2 nm

band at 465 nm belongs to the exciplex. The following values of Δv were obtained for a number of aromatic hydrocarbons quenched by diethylaniline (Table 10.1).

Although diethylaniline is able to donate an electron, the data from Table 10.1 are not in agreement with the model of reversible electron transfer. Actually, the S_i levels and the redox properties are different for various aromatic hydrocarbons [6, 499], however, the red shifts are centered at about 3000 cm^{-1} (the deviations are about a few hundreds cm^{-1}).

Table 10.1. Exciplexes of aromatic hydrocarbons with diethylaniline in *n*-hexane.

Aromatic Hydrocarbon	Q_m [%]	τ_m^0 [ns]	λ_m [nm]	v_m [cm^{-1}]	F_e [%]	λ_e [nm]	v_e [cm^{-1}]	Δv [cm^{-1}]
Anthracene	67	5.5	400	25000	2.0	465	21500	3500
3,4-benzpyrene	65	10.0	419	23900	6.3	472	21200	2700
Pyrene	61	10.0	392	25500	77.0	443	22600	2900
1,12-benzperylene	60	13.0	418	24000	36.8	471	21200	2800
Diphenyloxazole	60	3.4	371	27000	2.9	420	23800	3200
Azulene	50	–	378	26500	0.0	-	-	-
Chrysene	45	8.3	407	24600	2.5	470	21300	3300
1,4-di-(5-phenyl-2-oxazolylbenzene)	42	3.6	408	24500	1.0	470	21300	3200
Dimethylanthracene	25	75	420	23800	0.6	475	21100	2700
Coronene	1	20.0	450	22200	0.0	-	-	-

$Q_m=(1-F/F_0)$ is the quenching of intensity of the monomer fluorescence of the aromatic hydrocarbon after adding diethylaniline; the initial intensity (F_0) is 100%; F_e is the fluorescence intensity of exciplex; τ_m^0 is the initial lifetime of the monomer fluorescence, λ_m and λ_e are wavelengths of monomer and exciplex emission, respectively; v_m and v_e are the frequencies of monomer and exciplex emission

According to the IR spectroscopy data, diethylaniline has a broad and intense absorption band at 3000 cm^{-1} [236]. The extinction coefficient exceeds 200 M^{-1} m^{-1}, that is much greater than that of other aromatic hydrocarbons used. For diethylaniline in *n*-hexane the difference between the maximum of the excitation (or absorption) spectrum and the maximum of the emission spectrum is also about 3000 cm^{-1} (Fig. 10.3). Thus, we may assume that the high-frequency vibrations in diethylaniline when it accepts a part of the energy from the electronically excited donor are mainly responsible for the "red" shift in the exciplexes of aromatic hydrocarbons with diethylaniline.

10.4 Excimers

Concentration-dependent self-quenching of pyrene can be caused by vibrational combined IR transitions. The peaks in the absorption and luminescence bands of pyrene are represented in wavelengths and wave numbers scale (Table 10.2) [162]:

10.4 Excimers

Table 10.2. Maxima of pyrene bands.

	ν_0	ν_m	ν_e	ν_{vr}	$(\nu_m-\nu_e)$	$(\nu_m-\nu_{vr})$
nm	334	392	475	-	-	474[a]
cm^{-1}	29940	25510	21050	4430[a]	4460[a]	21080[a]

[a] the values calculated from the represented data.

The difference between ν_0 (the peak of the most intensive absorption band of pyrene) and ν_m (the maximum of luminescence of pyrene monomer) is proportional to the energy dissipated during vibrational relaxation (ν_{vr}) in the excited pyrene molecule. This energy does not significantly depend on the environment of the pyrene molecule. The position of absorption and luminescence bands of the pyrene monomer does not change with the polarity of the medium. Only the vibronic structure is changed [261, 262]. The value of ν_{vr} corresponds to the portion of energy that a pyrene molecule can accept when it acts as a quencher. The value of ν_{vr} reflects the average effective frequency of the IR transitions in a pyrene molecule. Therefore, the frequency shift between ν_m and ν_e, should be equal to ν_{vr}. In fact, the discrepancy does not exceed 30 cm^{-1}. In the range between 4600 and 4000 cm^{-1} pyrene has a number of combined frequencies (also a narrow peak at ~6 000 cm^{-1} is observed) (Fig. 10.4). In the near IR and visible regions pyrene is transparent. The position of the peak in the luminescence of the pyrene excimer, calculated as $\nu_m-\nu_{vr}$, coincides with the experimental value of ν_e. If this is not a random coincidence, then it should also be observed with other excimers.

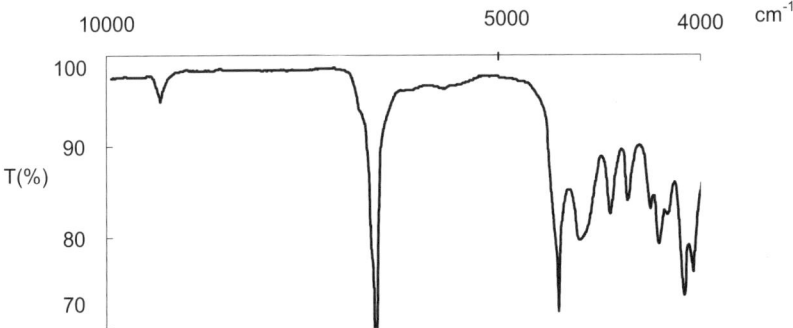

Fig. 10.4. Infrared spectrum of pyrene transmission in the region between 10000 and 4000 cm^{-1} [239]

The following values of $\Delta\nu$ were obtained for the excimers of aromatic hydrocarbons, in cm^{-1} [238]: 4200 for pyrene, 4100 for diphenyloxazole, 4300 for naphthalene, 4400 for benzene. Probably in these excimers not only the valency vibrations at 3000 cm^{-1} but also some other vibrations occurring at 1400–1000 cm^{-1} are effective [236, 237]. It leads to the formation of combined vibrations in the IR region at 4000–4700 cm^{-1} [237]. The finding that the difference between the positions of the peaks in the absorption (excitation) and emission bands of these hydrocarbons is much greater than that of diethylaniline also supports the viewpoint that combined vibrations are responsible for the «red» shift in the excimers. For instance, for pyrene (the absorption peak is at 334 nm, the emission peak is at 393 nm) this difference is about 4300 cm^{-1}, the value similar to the shift of the excimer band of pyrene relative to the emission of pyrene monomer.

It is well known that some aromatic hydrocarbons, such as pyrene, 3,4-benzopyrene, etc., have their own intermediate energy levels [53], while others do not. Thus, the transitions to the own intermediate levels of these donor molecules are also possible.

10.5 Pyrene–Indole Exciplex

Fig. 10.5 shows the emission spectra of pyrene in phospholipid liposomes before and after adding indole. Liposomes were used to achieve a high local concentration of the quencher. While quenching pyrene by indole, a broad low-intensity structureless band is observed at 460 nm. This band probably belongs to a pyrene-indole exciplex [146, 263].

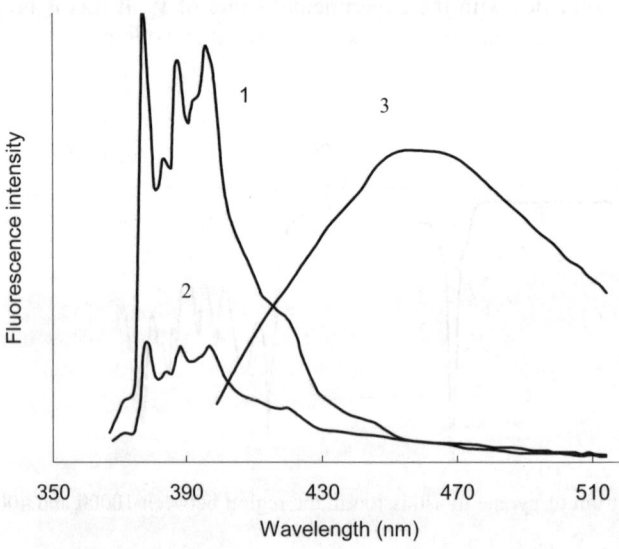

Fig. 10.5. Emission spectra of pyrene (1), pyrene after quenching and pyrene-indole exciplex (2) and the 15-fold magnified spectrum of a pyrene-indole exciplex (3), in liposomes. Concentration of indole is 0.5 mM [263]

10.5 Pyrene–Indole Exciplex

Table 10.3 summarizes the data on pyrene-indole and pyrene-diethylaniline exciplexes in liposomes and membranes of sarcoplasmic reticulum. It can be seen that the decrease in the intensity of pyrene fluorescence after addition of indole or diethylaniline is almost entirely due to a decrease in the lifetime. It should be emphasized that the spectrum of pyrene monomer fluorescence (with the maximum at 393 nm) does not overlap with the absorption bands of indole (280 nm) or diethylaniline (300 nm). Thus, Forster transfer of electronic excitation is impossible.

Table 10.3. Pyrene-indole and pyrene-diethylaniline exciplexes in liposomes and in sarcoplasmic reticulum [146, 263]

Membranes	Quencher	Fluorescence intensity [%]			τ_m^0 [ns]	τ_m [nm]	λ_e [nm]
		F/F_0	F_e/F_0	F_e/F			
Liposomes	DEA	45	6	13	152	79	450
Liposomes	Indole	25	6	24	152	40	460
SR	DEA	26	8	31	155	40	456
SR	Indole	49	5	10	155	81	–
SR+SDS	DEA	11	0	0	–	–	–

F_0 is the initial intensity of pyrene monomer luminescence (taken as 100%); F is the intensity after quenching; F_e is the intensity of exciplex emission; τ_m^0 is the initial lifetime of the pyrene monomer luminescence; τ_m is the lifetime after quenching; λ_e is the maximum of the exciplex emission. SR is sarcoplasmic reticulum; SDS is sodium dodecylsulfate; DEA is diethylaniline. The concentration of liposomes is 0.6 mg of lecithin/ml, the concentration of SR is 0.27 mg of protein/ml, the concentration of DEA is 300 µM, the concentration of indole is 500 µM, the concentration of pyrene is 0.4 µM, the concentration of SDS is 0.04%.

The fluorescence quantum yield for exciplexes of pyrene with indole or diethylaniline in membranes is small, tens time lower than in *n*-hexane.

Our findings are consistent with the results of Ruttens and coworkers [264] who demonstrated the exciplex formation in Na-acetyl-1-pyrenylalanyl-1-methyltryptophan methylester. When pyrene chromophore was photoexcited, a strong emission band of the pyrenyl-methyltryptophan exciplex was detected (in organic solvents). The excitation spectrum of the exciplex was almost identical with the absorption spectrum of the system. We may conclude that the excitation intensity at 280 nm belongs to the exciplex formed by an excited methyltryptophan and unexcited pyrenyl.

The formation of an exciplex between the excited tryptophan residue (or indole chromophore) and pyrene is possible. However, we were unable to record weak exciplex emission because of intense fluorescence from the protein as well as from pyrene monomers and excimers (Chapters 13, 14 and 16).

10.6 Conclusion

To explain the exciplex nature of tryptophan fluorescence, the exciplexes and excimers of heteroaromatic and aromatic hydrocarbons were studied. The values of the red shifts in exciplexes or excimers are centered at 3000 or 4200 cm^{-1}. A model of fractional energy transfer in exciplexes and excimers was proposed. The model suggests that the quencher accepts the vibrational quanta donated by the electronically-excited donor. According the model, the structureless red emission band appears due to the radiative transitions from the intermediate excited states of the donor. These intermediate levels are induced by the acceptor (in few cases, they belong to the donor).

The quenching of pyrene fluorescence by indole is observed. This quenching is due to deactivation and, probably, exciplex formation.

11 Mechanisms of Energy Transfer

The energy transfer processes could be classified as follows: (a) The energy transfer could be a short-range (distances of several Angstroms) and a long-range (distances up to several tens of Angstroms). In short distances the electronic clouds of the donor and acceptor are sufficiently overlapped. In long distances the overlap is small. The smaller is the donor-acceptor distance, the more efficient is the energy transfer. (b) The energy transfer is occurred due to strong or weak interactions. Under strong interactions between the donor in the ground state and the acceptor the absorption spectra are modified. Under weak interactions the absorption spectra are changed insignificantly. Usually, interaction between the excited donor and the acceptor is strong. (c) The energy transfer could be singlet-singlet, triplet-triplet, singlet-triplet, and triplet-singlet, depending on the spin state of the donor and acceptor. (d) The energy transfer could be «hot» and «cold». If the rate constant of the transfer is not less than that of the vibrational relaxation in the donor, this is a hot transfer. If the rate constant is less, the transfer is a cold. When transfer is hot, the sensitized and individual lifetimes are equal tacking into account the rate constant of vibrational relaxation of the donor. When energy transfer is cold, the lifetime of sensitized acceptor emission is approximately equal to the sum of acceptor own lifetime and the donor lifetime. (e) The energy transfer could be direct and reverse. If the electronic excitation transferred to the acceptor returns to the donor, this is a reverse transfer. (f) The energy transfer could be of Coulomb or exchange nature. (g) The energy transfer could be total (resonant) and fractional (non-resonant). If a quantum of energy is transferred from the donor as a whole, this is a total transfer induced by resonance between electronic levels of the acceptor and the excited donor. If only a part of the energy is transferred, this is a fractional transfer.

The main features of resonance energy transfer are: (a) the intensity and quantum yield of the donor luminescence is decreased in the presence of the acceptor; (b) the donor lifetime is decreased (it often accompanied by a non-exponential decay); (c) the shape of the donor emission spectrum remains unchanged in the course of quenching (if the donor luminescence is spectral homogeneous); (d) sensitized acceptor luminescence is developed, if the acceptor is able to fluoresce or phosphoresce; (e) a correlation between the intensity of sensitized acceptor and quenched donor luminesce should be observed; (f) the acceptor luminescence is depolarized (it may be not occur in the case of rigid fixation of parallely oriented chromophores); (g) in the excitation spectrum of the acceptor the intensity of the donor band is proportional to the transfer efficiency; (h) the efficiency of the energy transfer sharply depends on the donor-acceptor distance.

The resonance energy transfer in any biopolymer can be recognized with a set of all above features. One or two features are usually not enough to prove the resonant transfer, since each separate feature could be observed also in other processes. For instance, quenching of donor luminescence may be related not only to the transfer of energy of electronic excitation, but also to deactivation, electron transfer, etc.; shortening of the lifetime and non-exponential decay are observed in dynamic quenching as well; the emission spectrum could remain unchanged during screening and deactivation, etc.; acceptor luminescence may enhance due to reabsorption of donor emission; in thin layers the effect of volume reabsorption is possible; etc. To determine the efficiency of the resonant transfer in a certain system, all mentioned effects should be taken into account.

The energy transfer efficiency (E) is a ratio between the amount of quanta of electronic excitation which are transferred to the acceptor and the amount of photons absorbed by the donor. E is frequently calculated through the donor fluorescence intensity or lifetime in the presence and absence of acceptor, F or τ and F_0 or τ_0, respectively:

$$E = (F_0 - F)/F_0 \tag{11.1}$$

$$E = (\tau_0 - \tau)/\tau_0 \tag{11.2}$$

These formulae are suitable when quenching of the donor emission is caused only by electronic energy transfer. However, the acceptor molecule may quench not only by accepting of electronic excitation, but also by (a) direct deactivation (change a quantum of electronic excitation of the donor to a lot of vibrational quanta in acceptor); (b) electron transfer, (c) exciplex formation, etc.

11.1 Inductive-Resonance Model

Assume that the donor and acceptor are separated by a rather large distance so that their electronic absorption spectra are not deformed, i.e. the interaction is weak, and the electronic energy levels are not shifted. Assume also that the acceptor has an absorption band overlapping with the donor luminescence, i.e. the acceptor and the excited donor have a number of levels of equal of close energies. Thus, the effect of "resonance" is possible between the equal levels. This effect is similar to the mechanical resonance that is occurred between two pendula. The energy "flows" from one pendulum to the other. One quantum-mechanical oscillator induces a transition in the other. Therefore, a nonradiative transfer of excitation energy from the donor to the acceptor is feasible. Based on such approach, Forster developed a quantum-mechanical theory [147, 149, 265].

On the basis of perturbation theory Forster used the known equation for the probability of transitions between two states in terms of the quantum-mechanics (the "golden rule" [257, 266, 267]):

$$P = \frac{2\pi}{h} |\langle H \rangle|^2 \rho \tag{11.3}$$

where P is the probability of the energy transfer from the donor to the acceptor, $<H>$ is the matrix element of Hamiltonian of interaction, ρ is the density of the states. In the simplest case, $\rho=1/h\nu$ [126], where ν is the frequency of one mode of the acceptor. Thus, the lower is the frequency, the higher is the energy-transfer constant. Therefore, the energy transfer rate for a quantum of energy in the IR region is greater than those in the visible or UV regions.

$<H>$ could expressed as a sum of the Coulomb interactions between external electrons of the donor and acceptor and expanded in the Taylor series. Forster expressed the dipole-dipole approximation in the following equation:

$$H = \frac{e^2}{\varepsilon_0 R^3}\left\{(r_D r_A) - \frac{3}{R^2}(r_D R)(r_A R)\right\} \tag{11.4}$$

where R is the vector that connects two oscillators, r_D and r_A are the vectors that connects centers of the donor and acceptor molecules with electrons of their external electron shells, e is the charge of the electron, and ε_0 is the dielectric permeability of the medium at the optical frequencies. Forster [147] and Galanin [268] obtained the equation for the rate constant of dipole-dipole energy transfer. This equation expresses the wave functions in the adiabatic approximation through electron wave and vibrational functions and takes into account the absorption cross-section of the acceptor and the normalized spectrum of donor luminescence [147, 150, 268, 269]:

$$k_t = \frac{K^2 9000 \ln 10}{128 \pi^5 n^4 N \tau_0 R^6} \int_0^\infty F_D(\nu)\varepsilon_A(\nu)\frac{d\nu}{\nu^4} \tag{11.5}$$

where K^2 is the orientation factor, $F_D(\nu)$ is the spectral distribution of donor luminescence (normalized to the unit in the scale of wave numbers), $\varepsilon_A(\nu)$ is the molar extinction coefficient of the acceptor, N is Avogadro's number, n is the refractive index of solvent, τ_0 is the natural lifetime of the excited state of the donor in the absence of external quenchers, and R is the distance between the donor and acceptor.

(11.5) could be written as:

$$k_t = \tau^{-1}(R_0/R)^6 \tag{11.6}$$

where τ is the lifetime of the excited state of the donor in the absence of the acceptor. The parameter R_0 is called the "critical radius", i.e. the distance where the energy transfer to the acceptor and the spontaneous donor deactivation are equal.

Forster's equations (11.5), (11.6) establish the relationship between quantum-mechanical parameters, which cannot be directly measured (wave functions, Hamiltonian of interaction, density of states, etc.), and the measured parameters of the donor luminescence and the acceptor absorption. R_0 can be calculated from the known spectral characteristics of molecules. A theoretical k_t can be compared to the experimental results. The amplitude of the errors estimating the parameters K^2, Q_f, $\varepsilon_A(\nu)$, and n, changes a little the value of R_0. For instance, a 10-fold error in one of these parameters changes R_0 by less than 50%.

It was noted that Forster theory is an approximation [150, 162]. To elucidate the range where the Forster model could be applied, it is necessary to consider the assumptions that make the basis of this model. The perturbation theory used by Forster is based on the assumption that the value of $<H>$ is small relative to the Hamiltonian of an unperturbed system. In this case the Schrodinger equation is solved by expanding it in a series of $<H>$ degrees. The interaction energy is expressed as a series of different orders of the perturbation theory. For large distances a multipole approximation is used. Each consequent term in the series of the perturbation theory is expanded in a series of the inverse degrees of intermolecular distance. The use of only the second order of the perturbation theory is hardly strictly defined, since neglected terms could be of the same magnitude as the considered ones [267].

The expansion in a series of R degrees is valid only in the case when the distance is much longer than the radius of the electron cloud, $r_D << R >> r_A$. The parameter of convergence of the point-multipole expansion is the ratio of the length of the chromophores to the distance between them rather than the ratio of the length of the dipole moment to the distance. This is important for large organic chromophores with delocalized electronic clouds, of ~10 Å. The typical values of R_0 for organic molecules are ranging between 20 and 50 Å. Therefore, Forster's model can hardly be applied to quantitatively describe the resonance energy transfer.

The expansion of a function in a Taylor series, which is widely used in the perturbation theory, is a mathematical procedure, where each expansion hardly corresponds to a contain type of physical interactions [270]. The most complicated question is whether the expansion of any of the terms in the series of R degrees is valid. In chromophore molecules efficient energy transfer is occurred on short distances that are comparable to the sizes of the donor and the acceptor. In this case the expansion in the series of R degrees is incorrect because the error may be too large. The procedure is valid for large distances when $r_D << R >> r_A$. However, in this case also of the error could be large, since some of the terms in the series are omitted.

Forster's model uses only the dipole-dipole term in the series. Such an approximation is correct when the sizes of chromophore are considerably less than the distance. For example, when the chromophore sizes are 10 times smaller than the distance, the error of the dipole-dipole approximation is about 30% [271]. The error in the transfer rate is the same. When the distance between the chromophores is comparable to their sizes, the, error in the transfer rate constant may increase hundred times [162].

(11.5) has been derived using the expressions that relate the spectral parameters of the molecules, $F_D(v)$, $\varepsilon_D(v)$, and τ_0 to quantum-mechanical parameters, specifically, to the matrix element of the transition dipole moment, μ. These expressions are approximate, since simplified physical models of the quantum-mechanical oscillator and probabilities of optical transitions are used in both the quantum and classical theory of radiation. Therefore, the final Forster equation is of a qualitative rather than quantitative character.

The n^4 in (11.5) represents the refraction of the medium. It is derived from ε_0 in

(11.4). The relationship between n and ε_0 could be more complex; the ratio $n^2 \sim \varepsilon_0$ is approximately valid for optical frequencies. Furthermore, photoexcitation of the donor may result in changes in the local dielectric permeability of the medium surrounding the donor. Forster theory does not take it into account.

Forster theory assumes that the transfer rate should be significantly lower than the rate of the intramolecular vibrational relaxation of the donor. When this condition is violated, a hot transfer takes place (see below). This type of energy transfer cannot be described within the classical Forster theory. (11.5) could not be used when the rate of vibrational relaxation in the acceptor is too slow. When the rate is comparable to or lower than the transfer rate, a reverse transfer of energy to the donor may occur.

All the above do not imply the rejection of Forster theory. First, no better theory is available in this field at present. Second, the inductive resonance model is qualitatively agrees with some experimental data. In any case, this model qualitatively describes the processes observed: (a) with an increase in the distance between the donor and acceptor the transfer rate decreases, (b) with an increase in the dielectric constant of the medium the transfer efficiency decreases, (c) a high quantum yield of fluorescence of the donor provides a condition for a higher transfer rate, (d) a "favorable" chromophore orientation is required for more efficient transfer , (e) the quenching of fluorescence of the donor by the acceptor is stronger if the extinction coefficient of the acceptor in the region of donor emission is higher, (f) cold transfer is possible only when the emission and absorption spectra are substantial overlapped, etc.

The Forster model was modified many times but its essence and the key equation remained unchanged [150, 269, 272]. Dexter [266] developed the idea of inductive resonance and obtained the expressions in a more general form. The expressions, however, appear too bulky. Probably for that reason most of the researchers still prefer Forster's approach. A detailed information about various energy transfer models and experimental data can be found in [150, 162, 269, 272].

11.2 Energy Transfer in Molecular Structures

Models and experimental data related to the energy transfer in bichromophores, oligomers, polymers and biopolymers as well as in solutions of dyes and aromatic hydrocarbons are analyzed in [162]. A great number of accumulated data does not comply with the Forster model of inductive-resonance energy transfer over large distances. Many experiments carried out in the 50's and 60's as the evidence of long-distance transfer are rationalized now in a different way.

For instance, let us consider the well-known and widely cited work of Stryer and Haugland [273]. The authors state that they experimentally obtained the dependence of the transfer efficiency (E) on the distance predicted by Forster, i. e., $E \sim R^{-6}$. They used a bichromophoric molecule consisting of naphthyl, dansyl and an oligoproline chain between them.

However, the analysis of this work showed the following [162, 274]: (a) the excitation spectra of the donor and the acceptor were significantly changed with variations in the chain length; it indicates a strong interaction in the ground state, i.e. the perturbation was not weak; (b) when the transfer efficiency was 100%, the excitation spectrum differed from the absorption spectrum; moreover, the "donor" band in the excitation spectrum was even more intense; it which contradicts the energy conservation law; (c) the efficiency of transfer strictly depended on the excitation wavelength; it does not predict by Forster theory; (d) the size of the chromophores was approximately 7 Å and the value of R was varying between 15 and 46 Å; therefore, the dipole-dipole approximation is hardly applied; (e) if E measured at selected excitation wavelengths is plotted versus R, a set of absolutely different dependences but not the $E \sim R^{-6}$ dependence are obtained. Apparently, Stryer and Haugland just chose a "suitable" excitation wavelength.

Steady-state fluorescence was used in the study of the efficiency of singlet energy transfer from naphthalene to anthracene in bichromophoric diesters with values of m (the number of groups between the chromophores) in the range between 2 to 6 [275]. The dependence of the energy transfer efficiency on m agreed with Forster equation, the radius R_0 was equal to 16 Å. None of the compounds showed an exciplex red-shifted emission. The bichromophoric compounds had absorption spectra similar to that of an equimolar mixture of methyl 2-naphthoate and methyl 9-anthranoate. The authors assumed that the small band just above 400 nm was probably an artifact, because it was not observed in the excitation spectra. The spectra obtained in [275] revealed some non-additivity and "wings". Non-additivity and wings are the result of molecular interaction. Thus, the requirement of the Forster model on weak interactions were not met.

The fluorescence polarization properties of a pair of identical molecules electronically coupled were examined in [276] with stochastic Liouville formalism. Change of polarization in time was calculated for random ensembles of rigid molecular pairs under initial conditions that were represented either selective excitation or broad-band coherent excitation. The authors state that the Forster model could be applied to a strong coupling.

A theoretical and experimental study of the energy transfer in bifluorophoric molecules of terphenyl-CH_2-coumarin and naphthalene-CH_2-coumarin was conducted in [277]. The authors concluded that the mechanism of energy transfer did not comply with Forster's theory. They proposed an «inner-conversion» mechanism of energy transfer. The rate constant for this transfer depends on the overlap of the wave functions of the combined electronic-excited states.

An approximate solution for non-resonant transfer through a bridge described by a simple Huckel Hamiltonian was given in [278]. Exponential decrease in the coupling with the bridge length was predicted in all regions of the parameter space except proximity of chromophores and the resonance region. Resonance solutions were obtained when the levels of the donor and the acceptor lie either outside the occupied and virtual bands or within the band gap [278].

Many researchers in their studies of energy transfer did not quantitatively estimate screening, reabsorption, deactivation, exchange energy transfer, electron transfer, etc. These effects were analyzed in [60, 162].

Time-resolved fluorescence measurements of electron transfer in semi-rigid donor-acceptor compounds in viscous polar solution are represented in [279]. The results demonstrate a strong correlation between the intramolecular electron transfer rate and viscosity. The results were discussed in terms of dielectric relaxation dynamics of the solvent and viscosity-dependent conformational changes in the donor-acceptor system [279]. Time-resolved fluorescence measurements of intramolecular electron transfer in donor-acceptor compounds in various solvents were performed at different temperatures in [280]. Bi-exponential fluorescence decay was observed in moderately polar solvents and interpreted as evidence of intramolecular charge separation from the excited state to the ion-pair state and subsequent thermally activated repopulation of the excited initial state.

The processes of energy and electron transfer were observed in oligoproline bridged porphyrin donor-acceptor bichromophoric molecules [281]. Also, it was shown [282] that the main way in the decrease of donor fluorescence and phosphorescence of the cyclopentanporphyrin chemical dimers is the exchange energy transfer but not the resonance energy transfer. Photoinduced electron transfer was found in porphyrin-quinone cyclophanes [283].

Selective excitation of the two moieties of the dyad *Ru(II)-Rh(III) was carried out with visible and UV light in [284]. Electron- and energy-transfer processes were calculated with the use of transition absorption spectra. Both local excited states were quenched by electron transfer in liquid solutions at room temperature. No Ru(II)-*Rh(III) → *Ru(II)-Rh(III) energy transfer was observed, mainly because of a faster electron-transfer quenching. The electron transfer was blocked in rigid media or at 77 K and the energy transfer took place.

The processes of photoinduced electron and energy transfer in fatty-acid derivatives of diphenylpolyenes including stilbene, diphenylbutadiene, diphenylhexatriene, and heteroatom substituted stilbene derivatives were investigated in multilayer assemblies constructed with the Langmuir-Blodgett technique [285]. The absorption and emission spectra of the multilayer assembly were shifted to the shorter and longer wavelengths respectively. Absorption bands were strongly hypochromised. The majority of surfactant diphenylpolyene derivatives displayed spectral characteristics of "H" aggregates when incorporated into multilayer assemblies. Little if any energy migration occurred among the single stilbene-fatty acid monolayers, and enhanced energy migration ("an antenna effect") was observed in mixed multilayer assemblies. Although the fluorescence from the "H" aggregates could be quenched by energy transfer to an acceptor with a high oscillator strength, the energy transfer between stilbene aggregates was inefficient. Little or no "antenna effect" or interlayer energy transfer was observed between multilayer assemblies of the stilbene aggregate.

Absorption and emission spectra of donor-acceptor systems were studied in [286] using the general structure D-bridge-A, where the bridge consisted of an extended, rigid, saturated hydrocarbon skeleton that separated D and A by distances ranging from 3 to 12 C-C bonds. Sufficient electronic interaction induced a significant perturbation of the electronic absorption spectra across the bridges with a length of up to six bonds. Emission of a discrete charge-transfer type was detected in fluorescence spectra for bridge lengths of up to ten bonds. The magnitude of the electronic coupling matrix element ($<H>$) was exponentially decreased with the number of bridge bonds from 850 cm^{-1} at 3-bond separation up to 17.6 cm^{-1} at 10-bond separation. The rate of charge recombination in the studied compounds was proportional to the $<H^2>$ [276]. This result is the evidence of the "golden rule" – (11.3).

Other data on electron and singlet energy transfer in rigid supramolecular structures were considered in [162, 269, 287].

As objects for the study of energy transfer, biopolymers have sufficient advantages over solutions of dyes, aromatic hydrocarbons, rare-earth ions, and, in some respects, of bichromophoric molecules. The following facts confirm this statement: (a) In a biopolymer the donor of energy (tryptophan, nucleotide, covalent label) and the acceptor (coenzyme, substrate, label, probe) are in fixed positions at a certain distance from each other, whereas in molecular solutions chromophores are distributed chaotically. (b) A situation "1 donor - 1 acceptor" is often the case in a biopolymer whereas in molecular solutions a great number of donors and acceptors interact. (c) The proximity of the donor to the acceptor is achieved by their fixation at certain locations in a biopolymer. Therefore, there is no need for high concentrations, which result in screening and reabsorption of light in solutions of dyes and aromatic hydrocarbons. (d) When the donor and acceptor are separated by a large distance (tens of Angstroms), their diffusion-dependent collision is almost excluded in a rigid macromolecule, whereas in solutions the effect of diffusion is very important even at high viscosity. (e) The formation of sporadic donor-acceptor complexes may be avoided by incorporating a label or probe into a polymer, whereas in concentrated solutions such formation can hardly be eliminated. (f) X-ray diffraction data with a resolution of about 1 Å are available for nucleic acids and for some water-soluble proteins.

11.3 Hot Migration

In solution of molecules, their chromophores are collided on the nanosecond time scale. In macromolecules chromophores are arranged along the chain at distances comparable with chromophore sizes. The energy of interaction between chromophores in the chain varies from ~10 cm^{-1} to ~10000 cm^{-1} (the absorption spectra are distorted when energy of interaction is in the range between ~1000 and ~10000 cm^{-1}). In macromolecules chromophores can interact not only on the nanosecond but also on the picosecond time scale. Therefore, energy transfer from the unrelaxed ("hot") levels is possible.

A hot transfer from an unrelaxed vibrational state of a singlet excited state of the donor was proposed for crystals and solutions within the framework of semi-classical and quantum theory [288]. A method for calculation of the dependence of the probability of the hot transfer on the excitation frequency was developed under the assumption of strong electron-phonon interaction.

The hot transfer is possible only if interaction is strong enough and if the transfer rate exceeds the sum of all other rates of energy conversion.

We suggest that the probability of a hot migration between two identical chromophores can be expressed as the ratio of the area of absorption band located above the given excitation frequency to the area of the entire absorption band (Fig. 11.1):

$$P_{hot} = \frac{\int_0^{v_{ex}} \varepsilon(v)d(v)}{\int_0^{\infty} \varepsilon(v)d(v)} \qquad (11.7)$$

where P_{hot} is the probability of hot migration, ε is an extinction coefficient, v is the frequency in the absorption spectrum, v_{ex} is the frequency of the excitation light. This equation reflects four conditions: (a) the transfer is hot, it originates from the donor before vibrational relaxation, (b) any of the donor's quanta of energy with frequency greater than or equal to the lowest acceptor electronic levels can be transferred, (c) the chromophores have identical absorption spectra, (d) reverse transfer is absent.

The probability of hot migration depends on the excitation frequency [288]. Under "red-wavelength" excitation the number of electronic levels of the acceptor molecule with energy less than the electronic levels of the donor is minimal. The further the excitation wavelength is shifted to the red region, the lower is the probability of migration This can explain a drop in the migration observed with the red

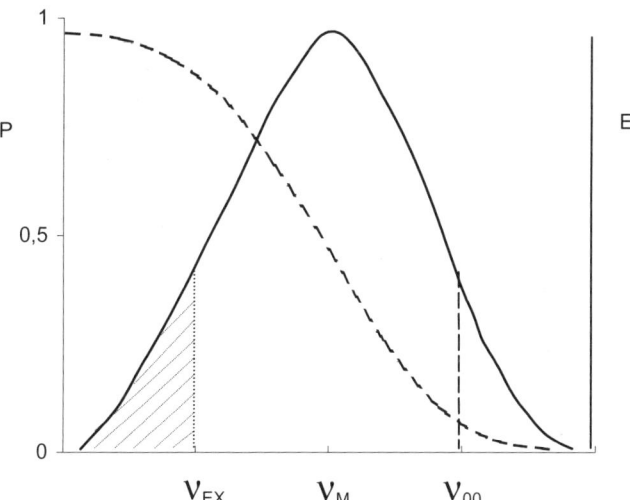

Fig. 11.1. Probability of hot energy transfer (----) between two identical chromophores as a function of the extinction coefficient of absorption band and the excitation frequency. v_m is the maximum in the spectrum, v_{00} is the 0-0 transition. Touched area in the absorption spectrum is the region where hot migration is absent

edge excitation (Weber's edge effect [7, 289]). It can be assumed that the "blue" part of the excitation spectrum during hot migration should be elevated since the hot migration competes with internal conversion.

Let us calculate the probability of the hot migration along an infinite chain of identical chromophores. If the probability of transfer from the first chromophore to the second one is P, the probability that the energy will be transferred to the third chromophore (the second step of migration) is:

$$P(2)=P^2 \quad (11.8)$$

For k-th step of migration:

$$P(k)=P^k \quad (11.9)$$

When $v_{ex}=v_m$, $P\sim0.5$ as follows from (11.7) and Fig. 11.1. Then $P(2) =0.25$, $P(3)=0.125$, etc. Thus, the migration becomes zero after a few steps. Therefore, in a polymer or biopolymer chain no efficient migration may occur along hundreds Angstroms. Efficient migration is possible only when (a) the "blue" edge of absorption spectrum is excited and when (b) the probability of a single migration step is close to 1.

It should be emphasized that a "cold" Forster migration between identical chromophores is forbidden because of the energy loss caused by the processes of vibronic relaxation.

11.4 Conclusion

(a) Strong overlap of the spectra of the donor and acceptor does not necessarily lead to the resonance transfer of electronic excitation. The quenching alone cannot prove energy transfer. The energy transfer efficiency could be estimated only comparing the value of the decrease in the donor fluorescence to the value of increase in the sensitized acceptor fluorescence.

(b) The Forster theory of inductive resonance focuses on the necessity of an overlap between the emission and absorption spectra for cold transfer, but fails to describe the phenomenon quantitatively, especially when the distance between chromophores is comparable to their sizes.

(c) No reliable data are available to prove the efficient transfer at the distances of 50-100 Å. Efficient transfer generally occurs when the chromophores are in close contact or separated by a distance of their own size.

d) Strong interactions may lead to the hot transfer. The probability of the hot transfer sharply decreases with decrease in the frequency of exciting light, that can explain Weber's edge effect.

(e) The cold energy migration cannot be efficient in macromolecules since it stops down after one or two transfer steps. The hot energy migration can be efficient.

12 Energy Transfer in Nucleic Acids

The possibility of electronic excitation energy transfer along nucleic acid chain (the energy migration) has been discussed in the literature for a long time [112, 162, 244, 290]. An answer to this problem is important both for understanding the role of migration in the UV-induced damage and reparation of DNA [38], and for using of energy transfer for fluorescence study of the conformational DNA behaviour [291].

The distances between nucleotides in DNA are about 3.5 Å, and planes of their chromophores are oriented stackwise, in parallel to each other. It preconditions efficient energy migration. The fundamental possibility of energy migration for long distances of hundreds of Angstroms along DNA was predicted in [112]. And, it was shown that the migration should be not "cold", but «hot», i.e. should occur not after the completion of vibronical relaxation, but in the course of it.

However, Rayner and coworkers [292] have found no migration at all along DNA from calf thymus, using the intercalating dye quinacrine as a «trap» of excitation. They have come to a conclusion, that energy is transferred only from four nearest to quinacrine bases (which surround the dye in the double helix). The conclusion about the absence of appreciable migration of electronic excitation along DNA was made. The phenomenology of energy migration in nucleotide units and crystals was studied in [293], but the mechanisms were not considered.

The discrepancy between the theoretical predictions and experimental evaluations of the efficiency of energy migration along DNA can be caused by the sharp decay of migration along the chain just after of one or two transfer steps, because of the low probability of each step.

12.1 Migration Between Nucleotides

The probability of one step of electronic excitation energy transfer between two adjacent chromophores can be expressed as:

$$P = K_t / (K_t + K_f + K_i) \qquad (12.1)$$

where K_t is the constant of energy transfer, K_f is the constant of the donor fluorescence, K_i is the deactivation constant, involving the processes of intersystem crossing, internal conversion, excimer formation, etc. For nucleotides in solution the relation $K_f / (K_f + K_i)$ is the quantum yield of fluorescence. At room temperatures it is very small. For example, for aqueous solution of adenine it is about 10^{-4} [117]. The low quantum yield of fluorescence of nucleotides makes the cold transfer by the inductive-resonant mechanism improbable. Even if the migration of electronic excitation along polynucleotides and nucleic acids is possible, it should be hot. For cold migration, internal conversion and external deactivation will predominate and migration will be impossible. Besides, the repeated transfer of excitation between identical or similar DNA chromophores by the mechanism of cold migration with weak overlapping of absorption and emission spectra can not be efficient. It is caused by the fact that vibronical relaxation in an excited nucleotide lowers the energy sharply just after the first step of transfer and makes the subsequent steps impossible. Only hot migration can be efficient.

K_f for nucleotides does not exceed 10^{10} sec^{-1} [112], that means, $K_i \sim 10^{14}$ sec^{-1}. In dinucleotides sufficiently strong chromophore interactions with a constant of the order of 10^{14} sec^{-1} causes a spectral shift of about 3000 cm^{-1} [294, 295]. For a nucleotide pair the probability of transfer will be close to 1, if $K_t \sim 10^{14}$ sec^{-1}. Such high constant corresponds to hot transfer [288, 296].

Strong interactions may occur between nucleotides. This can lead to spectral splitting, which especially pronounced in excitation spectra [293–295]. On the basis of the exciton theory it was calculated that the energy migrates between two nucleotides in a di- or polynucleotide within a time period of 10^{-14} s [295]. Not all nucleotides are able to strong interact. Among the strongly interacting nucleotides (adenine, cytosine), only a small portion of the population shows strong interaction. Therefore, efficient exciton migration along the chain seems to be unlikely in nucleic acids.

Rubin and Egupov [293] measured quenching of fluorescence and phosphorescence of adenine, guanine, cytosine and uracil aggregates by small amounts of 4-thiouracyl and explained it by singlet-singlet and triplet-triplet transfer of energy to the quencher.

Fluorescence quantum yields of individual bases and polynucleotides increases at low temperatures. It reaches 0.01–0.1 at 77 K [244]. Wilson and Callis [290] measured absorption and fluorescence spectra and polarized excitation and emission spectra of some nucleotides, their covalent dimers and polymers at low temperatures in ethylene glycol-water glasses. The polarization for dimers and polynucleotides was considerably lower than that of free bases of adenine and cytosine. For thymine and guanine, the polarization of dinucleotides did not differ from that for free nucleotides. A possible explanation is that both adenines and cytosines are capable of interacting with each other whereas thymines and guanines behave rather "independently". The authors observed a sharp increase in the fluorescence polarization in di- and polynucleotides when the red-edge excitation was used (Weber's effect. [7, 289]). Based on this finding, they concluded that neither Forster theory nor exciton theory seem appropriate here. Theoretical calculations for dinucleotides made on the basis of the exciton theory were found to disagree with the experimental data.

12.2 Migration Along DNA

The high constant of hot transfer is achieved in case of sufficiently strong chromophore interactions [288, 296]. If such interactions took place initially, before photoexcitation, they should show up in a shift and broadening of the absorption spectra. If strong chromophore interactions arise just after photoexcitation, spectra of absorption remain constant.

Maxima of absorption of nucleotides are known to be at 255–265 nm. The absorption spectrum of DNA has small shift, in any case, the spectral shift does not exceed 1–2 nm in comparison with the equivalent mixture of separate nucleotides; on the other hand, DNA spectrum is strongly hypochromized and broadened (Chapter 3).

In a chain comprising identical and equally oriented chromophores the probability of each transfer step is identical. If the probability of energy transfer from the first chromophore to the second one is P (the first step of migration), the probability of migration to the third chromophore (the second step) will be equal to P^2. For the k-th chromophore: $P_k = P^{k-1}$ (Chapter 11).

This migration proceeds without "jumps", since adjacent chromophores interact much stronger, than all the others. The probability of migration of an excitation along the chromophore chain drops, if $P<1$. For example, if $P=0.5$, the probability of four steps is only 0.0625, that is, migration practically ceases. Efficient migration along DNA could occur only on condition that P is close to 1.

It is extraordinary difficult to follow migration by the emission of bases for a number of reasons. Firstly, they emission is very poorly. Secondly, absorption and emission spectra of adenine, guanine, thymine and cytosine overlap strongly, i.e., it is difficult to distinguish adenine emission from that of guanine, etc. Thirdly, in polynucleotides and the majority of DNA species there is no a specific trap of excitation. And, lastly, strong emission of nucleotide excimers at room temperatures is observed [244]. Fluorescence of DNA is dominated by excimers involving adenine and thymine [244, 233]. Excimers arise at collision of adjacent nucleotides and these occurrences are limited by the conformational chain dynamics. In such conditions the migration of electronic excitation energy along the chain is hindered: excimers are "traps" in the migration circuit. At temperatures of about 300 K, polynucleotides form excimers very fast [244]. This is the other reason why efficient migration in polynucleotides cannot take place at room temperatures. Low temperatures allow to considerably decrease the excimer formation in polynucleotides [290]. However, even at 77 K the excimers could form, depending on the conditions of excitation and the type of oligo- or polynucleotide [294]. Even at 77 K, where the deactivation rate is decreased, it seems unlikely that energy migration will occur over a distance of tens of bases.

12.3 Quantum Yield of Energy Transfer to Dyes

For reliable migration recording a dye intercalated into DNA is necessary for transferring energy to it. If excitation is transferred to a dye, the quantum yield (efficiency) of migration Q_m can be expressed as a ratio of the quantity of electronic excitation quanta transferred to the dye molecules to the number of quanta absorbed by the polymer molecules. For the polymer labeled with a dye it can be written (at monochromatic excitation):

$$Q_m = F_{sa}/Q_a(I_d - I'_d) \qquad (12.2)$$

where F_{sa} is the intensity of sensitized acceptor fluorescence, i.e. the emission from the dye at donor excitation, Q_a is the fluorescence quantum yield of the dye at its direct excitation, I_d is the intensity of exciting light at the donor absorption band, I'_d is the remaining intensity of nonabsorbed light. F_{sa} varies along the excitation spectrum according to the form of the donor absorption spectrum (when Q_m=const and absorbance is small). The value Q_m at hot transfer should depend on the wavelength: it sharply falls at the "red-edge" excitation [288, 296].

According to the definition:

$$Q_a = F_a/(I_a - I'_a) \qquad (12.3)$$

where F_a is the fluorescence intensity of the acceptor at its direct excitation, I_a is the intensity of exciting light in the acceptor absorption band, I'_a is the remaining intensity of nonabsorbed light. F_a varies along the excitation spectrum and repeats the form of the acceptor absorption spectrum. From two above equations we obtain:

$$F_{sa}/F_a = Q_m(I_d - I'_d)/(I_a - I'_a) \qquad (12.4)$$

Using this expression provided the intensity of exciting light is the same at all wavelengths (the corrected excitation spectra) and applying the known Lambert-Beer law at small absorbances [138], it is possible to write:

$$Q_m = F_{sa}\, \varepsilon_a[a]/F_a\varepsilon_d[d] \qquad (12.5)$$

where ε_a is the extinction factor of the acceptor, $[a]$ is the acceptor concentration, ε_d is the average extinction factor of donor chromophores in a polymer, $[d]$ is the concentration of donor chromophores. Comparing the intensity of the donor band in the acceptor excitation spectrum with that in the absorption spectrum, it is possible to evaluate the migration efficiency, and the form of the band allows identifying hot migration. The obtained final equation is simple and exact, unlike too bulky and approximate equations in [292].

12.4 Polyadenilic Acid Labeled by Ethenoadenine

At room temperature the intensity of fluorescence of nucleotides and polynucleotides is very low (quantum yield is about 10^{-4}, lifetime is about 10^{-12}–10^{-13} s [244]). The efficient approach enables to work with polynucleotides at room temperatures is the introduction of special fluorescent labels into the polynucleotide. Excitation may "flow" to these labels. One of these labels is ethenoadenine, which is obtained by chemical modification of adenine. It has an additional absorption band at 300 nm and a fluorescence quantum yield of about 0.02. When few of adenine bases was modified in polyadenine chain, Razzhivin et al. [297] observed the adenine band at 265 nm in the excitation spectrum of the ethenoadenine label (the lable fluoresced at 410 nm). The more was a number of modified bases, the lower was the intensity of adenine band. The intensity of adenine band was also decreased, when the solution was heated. It was concluded that efficient energy transfer from adenine bases to an ethenoadenine label takes place and that its efficiency is decreased when the interactions in the polymer are destabilized.

Let us try to evaluate the efficiency of energy migration in polyadenilic acid using the experimental data from [297]. The authors have observed the energy transfer from adenines of poly-A to the fluoresce ethenoadenine label, attached to the polymer. At 300 nm the ε of ethenoadenine is equal to 3000 $M^{-1}cm^{-1}$ [297], and that of adenine chromophore in poly-A at 265 in, is known, to be about 10000 $M^{-1}cm^{-1}$. It means that $\varepsilon_a/\varepsilon_d = 0.3$. This value is close to the ratio of ethenoadenine fluorescence intensities at 300 and 265 nm excitation, ~0.3, which was observed in the excitation spectrum of the polymer with low degree of updating, 0.035. It appears, that energy is transferred to the label only from one adenine chromophore, probably nearest to ethenoadenine. Using the final equation at $[a]/[d]=0.035$ we obtain that $Q_m=0.032$ (Table 12.1), i.e., approximately only 3 quanta of electronic excitation reach the label of each 100 quanta absorbed by the molecules of polyadenilic acid. Thus, at room temperatures electronic excitation energy cannot migrate over long distances along polyadenine.

12.5 DNA with Intercalated Dyes

In natural nucleic acid molecules the situation should be more favorable for energy migration, especially in phage DNA. Phage DNA is sufficiently rigid structure even at room temperature. Therefore, the deactivation processes (in particular, the formation of excimers between the bases) is hindered, and also, chromophore planes are favorably oriented in parallel to each other. It creates preconditions for migration. The low fluorescence quantum yield of nucleotides and weak interference of their absorption spectra allow the conclusion, that the migration between DNA bases can be only hot.

One of a convenient acceptor of the energy from bases is ethidium bromide (EB). EB is an intercalating dye [184, 298, 299–302] with intensive absorption

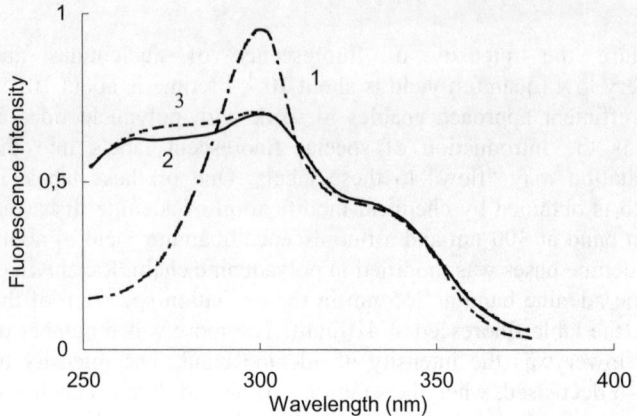

Fig. 12.1. The excitation spectra of ethanol solution of ethidium bromide (1) and that bound to native DNA (2) and heat-denaturated DNA (3) from phage «Lambda» in 10 mM tris-HCl, pH 7.6 at 20 °C. The concentration of the DNA bases is ~18 μM, EB is 1.5 μM. The EB emission was recorded at 630 nm [117]

bands at 300 and 335 nm and with high Q_a. EB is distributed rather uniformly along DNA [118]. Such properties create preconditions for the excitation transfer to EB, as well as permit separate observation of F_a and F_{sa}. In model experiment, F.Van Nostrand and R.M.Pearlstein [303] observed that the fluorescence of poly rA frozen in an aqueous medium containing sugar was half-quenched by ethidium at $[a]/[d]$=0.01. Unfortunately, they did not study the mechanism of quenching.

In Figure 12.1 the excitation spectrum of fluorescence of EB bounded to DNA from phage "Lambda" in the ratio $[a]/[d]$~0.12 is presented [117]. At such ratio the dye molecules are located close to each other and separated by about six nucleotide pairs only. The maximum of the excitation spectrum of EB on DNA is at 300 nm, and there is a shoulder at 335 nm. Such spectrum is similar to the spectrum of EB in ethanol, but unlike it has an intense additional band at 260 nm. Besides, the 300-nm band of EB on DNA is as if "cut off" due to hypochromism (Chapter 3). In the aqueous solution of the dye (not shown) this band is shifted to 286 nm, and the shoulder is not pronounced.

Some authors supposed that the shoulder at 335 nm in the spectrum of EB bound to DNA to be a result of exciton splitting on interaction of adjacent dye

Table 12.1. The efficiency of energy migration along polynucleotide chains with excitation transfer to ethenoadenine and ethidium [117]

Donors	Acceptor	$[a]/[d]$	Q_m	D/A
Adenines of poly-A	Ethenoadenine	0.035	0.032	~1
Bases of «Lambda»	Ethidium	0.12	0.15	1.25
Bases of «T7» DNA	Ethidium	0.01	0.02	2.0

12.5 DNA with Intercalated Dyes

molecules. However it was «... not clear, how the intercalated EB molecules could interact each other with more than one pair of bases are them; the possibility of chromophore-chromophore interactions under such conditions (a distance of 10.4 Å between the planes of EB chromophores) seems to be much more doubtful" [299]. As it is seen from Figure 12.1, there is the same pronounced band in the spectrum of EB in ethanol solution. It is caused by usual chromophore-solvate molecular interactions and is not excitonic.

The 260-nm band in excitation spectrum (Fig. 12.1) belongs to nucleotides and is due to migration from them to EB. The ratio of the intensity of the DNA band at 260 nm to the one of the EB band at 300 nm is about 0.8, The ε of DNA at 260 nm is about 8350 $M^{-1}cm^{-1}$ per one nucleotide. The ε of EB on DNA in the band maximum is 13000 $M^{-1}cm^{-1}$. The ratio of the ε values is equal to 0.64. From the ratio 0.8/0.64 the average number of nucleotides donating the excitation to one dye molecule (D / A)=1.25 is obtained. One or sometimes two nucleotides participate in the excitation transfer. Using the final equation at $[a]/[d]$=0.12 we obtain Q_m=0.15 (Table 12.1). Only 15 excitation quanta of each 100 light quanta absorbed by the nucleotides are transferred to the dye molecules. The Q_m value did not depend on the nucleotide excitation wavelength, i.e. hot migration between the nucleotides is not found.

Is not it EB that influences the efficiency of migration along DNA? Unwinding of DNA chains is known to be induced by large concentration of EB [300, 301]. Unwinding of DNA can be also caused by quinacrine and other intercalating dyes [300]. As it is seen from Figure 12.1, thermal denaturation of DNA-EB complex does not influence the excitation spectrum. It suggest that either binding of the mentioned EB quantities caused DNA denaturation, or (and) the transfer of excitation to the dye is not sensitive to disruption of the DNA structure.

Actually, with EB binding to DNA, hyperchromism of DNA in 260-nm region is observed (not shown). It is due to DNA denaturation, causing the removal of hypochromism of native DNA. Thus, this EB concentration does induce DNA denaturation. Thermal denaturation of the DNA-EB complex did not influence the differential absorption spectrum. Hence, the transfer of excitation to EB at the mentioned dye concentration occurred under the conditions of strong EB-initiated infringement of the native DNA structure.

The less is the ratio $[a]/[d]$, the less should be denaturation induced by the dye. Contribution of the DNA 260-nm band in the excitation spectrum increased appreciably as the EB concentration reduced. It is shown in Figure 12.2 (with DNA from phage «T7»). For the small dye quantity, when $[a]/[d]$=0.01, ratio F_{sa}/F_a makes up more than 1.3. Substitution of the experimental data in the final equation shows, that here Q_m=0.02, i.e. it is less than at the large dye concentration. Only two excitation quanta of 100 quanta absorbed by the DNA bases reach the dye molecules. This result is determined by the fact that with $[a]/[d]$=0.01 about 50 pairs of bases are between the EB molecules. The majority of nucleotides are far from the dye and they cannot donate the energy of electronic excitation. The Q_m value did not depend on the wavelength of excitation of nucleotides, i.e. hot migration between nucleotides is not found.

F_{sa}/F_a exceeded ~2 times the $\varepsilon_d/\varepsilon_a$ ratio. Therefore, only two nucleotides can be energy donors for one dye molecule, $D/A=2.0$ (Table 12.1). This must be the two closest (to the dye) nucleotides. We note also, that at $[a]/[d]=0.01$ energy migration between the dye molecules was excluded completely.

At small quantities of the dye, DNA does not untwist, but local disruption of

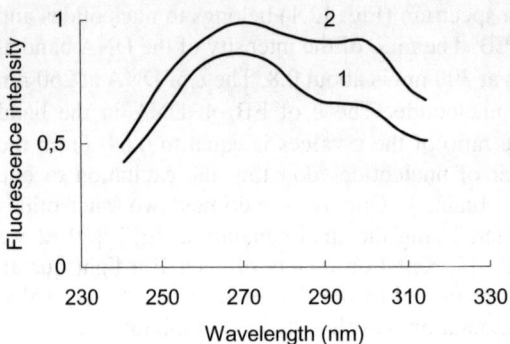

Fig. 12.2. The corrected excitation spectra of ethidium bromide on DNA from phage «T7» at [EB]/[DNA bases]~0.01 (1) and at [EB]/[DNA bases]~0.24 (2). Excitation slits are 6 nm; emission is 630 nm. The intensity of the second spectrum scaled down by a factor of 20

the DNA structure in the places of intercalation can arise. It could cause migration degradation through disruption of stackwise arrangement of bases near the dye. Therefore, it is appropriate to plot F_{sa}/F_a versus the EB quantity and to approxima-

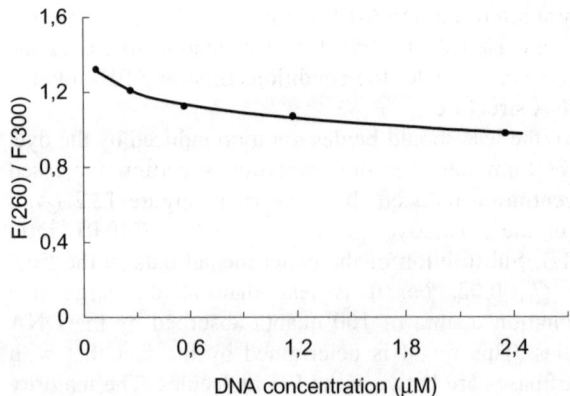

Fig. 12.3. Ratio of intensities at 260 and 300 nm in excitation spectrum of ethidium bromide bonded to DNA from phage «T7» versus dye concentration. The concentration of DNA bases is ~10 µM. Emission is at 630 nm

te it to zero concentration (Fig. 12.3). When approximated to zero dye concentration, the F_{sa}/F_a ratio reaches 1.44; i.e. it differs little from F_{sa}/F_a at $[a]/[d]=0.01$. It means, that the dye in small concentration does not disrupt the DNA structure, and hence does not influence the migration between nucleotides in any possible way.

12.6 Fluorescent Probes and Labels on DNA

The fluorescence of anthryl chromophore was efficiently quenched by bases when anthryl was binding to synthetic and natural DNA [234]. Time-resolved fluorescence measurements clearly show biexponential decay behavior (8.2 and 30.6 ns) for the anthryl chromophore bound to cytosine-thymine DNA. Triplet-triplet absorption spectra of the probe in the presence of cytosine-thymine DNA show complete quenching of the anthryl triplet by the DNA bases. Excitation into the absorption bands of DNA in the region between 260 and 300 nm (where the absorption of anthryl is small) resulted in an intense, red-shifted, and broad fluorescence spectrum of the anthryl chromophore. When the helix was melted, energy transfer could no longer be observed. Thus, the double-helical structure of the DNA polymer is important for the energy transfer. The excitation spectrum of the probe in the presence of cytosine-thymine DNA showed a new band in the region 260–300 nm. The spectrum of sensitized emission was quite similar to the time-resolved emission spectrum of the long-lived component. The energies of the singlet excited states of all the four DNA bases are much higher than that of anthracene; hence, singlet energy transfer from the bases to the anthryl group should be exothermic [304]. Energy transfer from DNA to pyrene or to perylene chromophores was not observed, although energy transfer to these chromophores was expected to be also exothermic. While the intensity of the sensitized emission decreased with the helix melt, the intensity of the direct emission increased. The efficiency of sensitization decreased with the increase in the guanine-cytosine content. No energy transfer was observed with poly(dG-dC). The maximum efficiency of sensitization was observed at 285 nm, near the red edge of the absorption spectrum of DNA. Absorption of light by DNA at shorter wavelengths did not lead to sensitized emission [304].

In the absence of DNA, zirconium phosphate with butylamine hydrochloride was found to enhance singlet-singlet energy transfer from 1-naphthalenemethylamine hydrochloride to 9-anthracenemethylamine hydrochloride or to 3-(l-pyrenyl)butylamine hydrochloride [348] (theoretical $R_0=25$–30 Å; the emission spectra of the donor and the acceptor were measured as a function of the acceptor concentration). No energy transfer was observed when these probes were bound to DNA or to proteins [305].

The authors of [306] used a luminescent europium chelate as a donor and an organic dye (CY-5) as an acceptor. A double-stranded DNA oligomer served as a rigid structure to set a definite distance between europium and CY-5. The donor and acceptor were separated by 42 Å. However, there was some uncertainty in the exact position of the dyes because of the flexible six-carbon linkers used for attachment. Calculated R_0 was equal to 70 Å in D_2O and 56 Å in H_2O. The luminescence of the donor without the acceptor was single exponential with lifetime of 2.52 ms. The luminescence of the donor in the presence of the acceptor was bi-exponential (0.22-ms component, 63%, and 2.40-ms component, 37%). It was assumed that the fast component was the result of intramolecular energy transfer in the hybridized donor-acceptor complex at 91% quenching. A titration with increasing acceptor concentration increased the amplitude of the fast component but did not change its lifetime. The sensitized luminescence of the acceptor was bi-exponential (0.059 ms, 40%, and 0.25 ms, 60%). The fast component was due to direct excitation of the acceptor. The authors assumed that the 0.25-ms component as well as the fast component of the donor quenching were due to energy transfer. Unfortunately, the decay of the acceptor without the donor was not shown in the Figures in the paper, i.e. the control experiment was missing.

Resonance energy transfer between fluorescein and rhodamine covalently attached to double-stranded DNA species was suggested in [307]. Experimental data were interpreted according to Forster theory. Unfortunately, no any spectra are shown in the paper for confirmation of the interpretation.

The binding of tetramethylrhodamine-5'-(GGGCTATAAAAGGG)duplex-3'-fluorescein by native TATA binding protein were investigated in [520]. The protein addition led to quenching of the fluorescein emission. The quenching was interpreted as result of increase in resonance energy transfer efficiency from fluorescein to rhodamine due to conformational transition in the duplex. However, no any control spectral data without rhodamine (potential acceptor) are represented in the paper. Moreover, in the duplex without rhodamine the value of the short-lived fluorescein lifetime component was decreased by ~30% upon adding the protein, i.e., the protein directly influenced on the fluorescein emission.

A method for the rapid and direct identification of a single point mutation in a DNA sequence using energy transfer was developed in [308]. A 16-base oligomer with 3'-bound fluorescein (donor) and 5'-bound x-rhodamine (acceptor) serves as a probe. A dramatic difference in the probe emission in duplex structures and as a single strand was detected This difference was used to obtain the melting temperature of transition from double to single strand. Unfortunately, the authors did not represent the control data on donor emission without the acceptor and acceptor emission without the donor. Therefore, the efficiency of resonance energy transfer is indefinite.

Using near-field scanning optical microscopy (NSOM), the authors of [309] have measured the single-molecule fluorescence of tetramethylrhodamine and Texas Red attached to the 5' ends of synthetic DNA oligomers placed on the metallic coating of the NSOM tip. The efficiency of energy transfer was estimated using photobleaching as the method to destruct the donor or acceptor. For the oligomer of length 10 bases, a strong interaction between the dyes was found. For the oligomer of length 20 bases, too high efficiency, 85%, was obtained. These strange facts were explained flexible linkers bringing the dyes as close as possible. The authors did not synonymously resolve the doubt: do the detected phenomena involve resonance energy transfer or not? On the other hand, the obtained color NSOM images of labeled single DNA molecules may be successfully applied to detection of the DNA hybridization, melting, etc. A single-molecule fluorescence of tetramethylrhodamine and Texas Red attached to a short single strand DNA adsorbed on a silianized glass surface was studied in a millisecond time scale in the far field with linearly polarized excitation light which was continuously modulated [310]. A rotational jump of a single dipole, photobleaching, a spectral jump, and desorption from the surface were detected.

In [311] energy transfer was used to detect intracellular oligonucleotide hybridization.

12.7 Conclusion

The efficiency of hot energy migration along DNA with passing of excitation to fluorescent dye is considered. The efficiency of energy migration along DNA is low. Using intercalated dyes it was shown, that at room temperatures the hot energy transfer to dye occurs only from 1–2 neighbouring nucleotides. The divergence received experimental data from literature calculations about possibility of effective energy migration along DNA nucleotides is caused by recession of migration already afterwards of one–two transfer steps because low probability of each step. At room temperature electronic excitation energy cannot migrate over long distances along nucleic acids. Probably, the efficiency may be increased in inside-phage DNA or in the isolated DNA that immobilized, frozen, or crystallized. The migration should be hot.

13 Energy Transfer in Native Proteins

For several reasons it seems unlikely that the peptide bonds in proteins can act as transmitters of electronic excitation under normal conditions. First, electronic transitions should occur at very high frequencies, i.e. in the UV region at 190 nm, and the fluorescence quantum yield is almost zero. Second, peptide bonds have relatively intense vibrations in the IR region. Therefore, very fast internal conversion should take place. Third, peptide bonds in proteins are surrounded by amino acid residues, which can efficiently deactivate the excited peptide bonds.

Aromatic amino acid residues: tryptophan, tyrosine and phenylalanine as well as special chromophore prosthetic groups and coenzymes (for many redox proteins and enzymes): flavins, hemes, nicotine-amid-adeninenucleotides, and retinals are potential donors and acceptors of electronic excitation energy in protein.

The fluorescence of tryptophan and tyrosine is usually strongly quenched in proteins containing prosthetic chromophore groups or in enzymes bound to substrates or coenzymes containing a chromophore. These are heme proteins (cytochromes, globins, catalases, peroxidases), retinal-containing proteins (rhodopsin, bacteriorhodopsin), copper-proteins (azurin etc.), NADH-dependent dehydrogenases, flavine oxidases, luciferases, etc. Effective energy transfer in these systems usually occurs over distances of about 10 Å [60].

13.1 Tyrosine–Tryptophan Pair

The efficiency of energy transfer from tyrosine to tryptophan residues is very interesting to be studied [60, 67, 15–18, 312]. A theoretical value of the Forster radius for the pair tyrosine - tryptophan is about 10–14 Å [67, 17, 18, 313, 314]. Calculations were based on the quantum yield of fluorescence of tyrosine and the integral of the overlap between the emission spectrum of tyrosine and absorption spectrum of tryptophan in aqueous solutions of these amino acids.

However, spectral parameters are quite different in proteins: the quantum yield of tyrosine residue is substantially lower than that of a free tyrosine; the absorption spectrum of free tryptophan in water is usually wider than that of tryptophan residues; the peak of the absorption band of free tryptophan in water is red-shifted by several nanometers and varies from one protein to another. It should be noted that a low quantum yield of tyrosine fluorescence is also registered in proteins containing no tryptophan residues [17, 313]. The exact estimation of R_0 for aromatic residues in proteins seems hardly possible even when the orientations are available from X-ray diffraction data.

The efficiency of energy transfer from tyrosine to tryptophan residues was estimated in different ways by various authors. Some of them interpret their results for the same proteins as indicative of efficient transfer, others deny any presence of it. The following are some examples.

Kronman and Holmes [164] measured quantum yields of fluorescence from a number of proteins using two excitation wavelengths: 295 nm, where tryptophan residues are preferentially excited, and 280 nm, where both tyrosine and tryptophan residues are excited. They estimated the efficiency of energy transfer from tyrosines to be 100% in trypsin, 50% in papain, 80% in pepsin and >45% in albumin. Lerner et al. registered the efficiency of energy transfer as only 27% in trypsin [315] that was similar to the efficiency of 29% determined in the experiments of Saito et al. [316]. No energy transfer from tyrosine to tryptophan residues in lysozyme, chymotrypsin or chymotrypsinogen was revealed in [315]. Based on the Forster theory, Turoverov et al. [317] predicted 50% efficiency of energy transfer from one of the two tyrosine residues in azurin to the single tryptophan residue in this protein. The authors used X-ray diffraction data for azurin crystals with an exact determination of the distance between the chromophores and their mutual orientations. However, no energy transfer was observed in the experiments [243].

Using Forster theory and X-ray diffraction data for γ-crystalline, Borkman et al. [314] predicted the efficiency of energy transfer of about 83% from 15 tyrosine residues to 4 tryptophan residues. Their experiments yielded the value of 78%. Such good agreement can be explained by various adjustments made by the authors because neither parameters of tyrosine fluorescence in the absence of tryptophan residues nor fluorescence and absorption spectra for individual residues are known for γ-crystalline; reliable data on the overlap of the spectra are completely lacking.

Experiments on ketosteroid isomerase mutants using time-resolved fluorescence and fluorescence anisotropy demonstrated that in a hydrophobic environment Tyr-14 had an unusually long fluorescence lifetime (4.6 ns) as compared to Tyr-55 (2.0 ns) or Tyr-88 (0.8 ns) and to most tyrosine residues in proteins (0.2–2 ns) [318]. Forster radii obtained from the absorption and emission of these tyrosines predict that a total quenching of Tyr-14 fluorescence by Tyr-55, and to a lesser degree by Tyr-88, would take place if their orientations were favorable. The distances calculated using the Forster equation (11.6) are 12.5 Å between Tyr-14 and Tyr-55, 11.1 Å between Tyr-14 and Tyr-88, 10.4 Å between Tyr-55 and Tyr-14, etc. The authors came to the conclusion that the lack of efficient quenching of Tyr-14 by Tyr-55 implies that Tyr-14 and Tyr-55 are oriented unfavorably, and that this orientation is rigid on the time scale of picoseconds to nanoseconds. The rigidity of Tyr-14 and Tyr-55 was confirmed by time-resolved fluorescence anisotropy at 20° and 40°C, where the only correlation time corresponding to the global motion of the protein was resolved. Three factors probably account for the unusually high fluorescence of Tyr-14 [318]: (a) the hydrophobic environment; (b) the absence of quenching groups nearby; (c) the rigidity of Tyr-14 environment.

The 30% efficiency of energy transfer from Tyr to Trp in the peptide hormone adrenocorticotropin was reported in [312]. Unfortunately, the statement was not confirmed by spectral dependence of the Trp fluorescence quantum yield on the excitation wavelength with correction to the contributions from the Tyr and Trp absorption.

Excitation, absorption and emission spectra, quantum yield, and lifetime were measured using various excitation wavelengths between 260 nm and 300 nm for a number of water-soluble proteins as bovine serum albumin, lysozyme, trypsin, chymotrypsinogen A, pepsin, alcohol dehydrogenase, subtilisin, hyaluronidase, hexokinase, as well as the polypeptide glucagon and DL-tryptophan [60]. Narrow monochromator slits were used (1–2 nm), the scanning step was 1 nm and precise measurement of absorbance was provided. The experiments revealed less than 10% efficiency of energy transfer from tyrosine to tryptophan residues. Albumin was the only exception, where the efficiency was about 30%. An error in the estimation of the energy-transfer efficiency could be due to hypochromism of aromatic residues (Chapters 2 and 3). The data on the energy-transfer efficiencies are summarized in Table 13.1.

No tyrosine-tryptophan energy transfer was found in phospholipase A2 or ribonuclease T1 using excitation spectra [119]. This can be due to: first, a high hypochromism (non-additivity of extinction coefficients of chromophores in these proteins), and, second, absence of tyrosine-tryptophan interactions, when the distances between the chromophores are more than 5 Å.

Bovine phospholipase A2 is a small protein containing 7 tyrosine residues separated by short distances from a single tryptophan. Estimation of the inductive resonance transfer by the Forster model (using X-ray diffraction data to determine distances and orientations in phospholipase A2) indicated that most of the tyrosine

Table 13.1. Efficiency of the energy transfer from tyrosine residues to tryptophan residues in native proteins

Protein	Efficiency, [%]
Bovine serum albumin	<30
Lysozyme	<10
Trypsin	<10
Chymotrypsinogen A	0
Pepsin	<10
Alcohol dehydrogenase	0
Subtilisin	<10
Hyaluronidase	0
Hexokinase	0
Glucagon	<15
Phospholipase A2	0
Ribonuclease T1	0

Experiments were carried out with "SLM-4800" in the 100X averaging mode. Absorbance is approximately 0.1; the accuracy of spectrophotometric measurement was 0.005; the accuracy of extinction coefficients for tyrosine and tryptophan residues was approximately 10% [60].

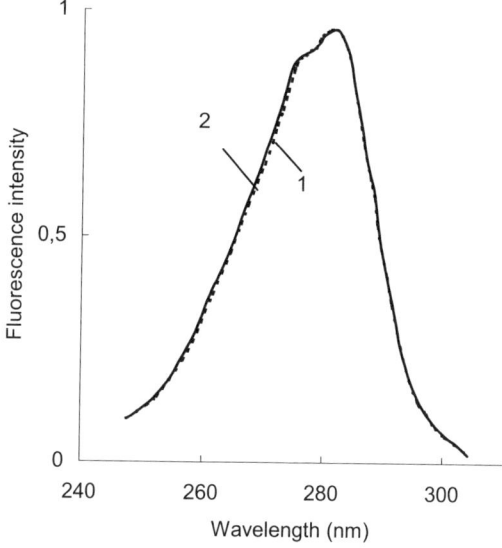

Fig. 13.1. Excitation spectra (non-corrected) of bovine phospholipase A2 (1) and NATA (2) in acetate buffer, pH=6.0. Emission is at 350 nm; excitation slits are 4 nm; "Perkin-Elmer MPF 44B"

residues should be good donors of energy to tryptophan; the theoretical average efficiency of energy transfer should be ~70%. However, no significant energy transfer was revealed in the experiments. Fig. 13.1 shows normalized excitation spectra of bovine phospholipase A2 and NATA. The spectra coincide; no tyrosine component in the spectrum of phospholipase A2 is visible [119]. It may be assumed that the seven tyrosines in bovine phospholipase are quenched by some non-chromophore groups rather than by the tryptophan residue [119]. This quenching does not disappear in viscous glycerol solutions (Chapter 9). Therefore, it is not dynamic quenching limited by flexibility of a protein globule; probably, it is static quenching.

Similar to the situation with phospholipase A2 and ribonuclease Tl, in most proteins tyrosine fluorescence is quenched by various non-chromophore groups to such an extent that tryptophan residues cannot compete successfully for the excited tyrosines. Efficient energy transfer can take place only when the tyrosine and tryptophan are arranged so that they can contact each other and there are no other quenching groups in their surroundings.

It seems that some authors overestimated the accuracy of the methods used in the calculation of the tyrosine-tryptophan energy transfer efficiency. In many proteins with no tryptophan residues tyrosine fluorescence is often strongly quenched by carboxyl- and amino-groups [67, 17, 18].

13.2 Migration between Tryptophan Residues

Absorption spectrum of tryptophan in solution is slightly overlapped with the fluorescence spectrum. The calculated Forster radius, R_0, is equal to about 6 Å [15, 17, 313]. Since the spectral parameters of tryptophans differ depending on the protein, the calculations should be made for each individual protein. Attempts have been made to determine experimentally the efficiency of energy migration between tryptophan residues in proteins by fluorescence quantum yield or by fluorescence depolarization. Some investigators have used time-resolved fluorescence spectroscopy for this purpose. As an example, the work of Desie et al. [319] will be considered below in more detail.

The kinetics of the decay of tryptophan fluorescence in crystalline chymotrypsin was triple-exponential; the decay times being 0.15, 1.45 and 4.2 ns. This allowed the authors to decompose the emission spectrum in three components. The maxima of the bands were at 325, 332 and 343 nm. Based on the X-ray analysis data, the authors referred these bands to three tryptophan-emitting centers. These are: (a) Trp-172 and Trp-215; (b) Trp-51 and Trp-237; (c) Trp-27, Trp-29, Trp-141 and Trp-207. The tryptophan residues in each center are localized in the vicinity of each other. They were grouped with respect to both the distance between the chromophores and their mutual orientation. The efficiency of energy migration between these tryptophans was calculated using the Forster formalism. The authors came to the conclusion that the energy transfer efficiency within each group is very high but the energy transfer between three different centers is absent. The tryptophan residues were divided into groups in accordance with three spectral components as follows. The nitrogen on the indole ring of Trp-172 forms a hydrogen bond with the carbonyl group of proline-225, and Trp-215 forms a hydrogen bond with methionine-180. Moreover, at a distance of 5.5–6.1 Å from these tryptophans there are also other residues capable of quenching fluorescence. Thus, the authors concluded that the emissions of tryptophans in this centers should be quenched very strongly, i.e. the lifetime of this center is of 0.15 ns. Similarly, the second and the third centers were associated with lifetimes of 1.45 and 4.2 ns with the bands at 332 and 343 nm, respectively.

The following remarks can be made here.

(a) The detection of the three components of fluorescence decay of chymotrypsin was rather qualitative; the method used by the authors was unable to precisely differ similar lifetimes, for example, 4.2 ns and 3.9 ns. Indeed, according to the authors, the experimental error was 0.05, 0.25 and 0.4 ns for the three components, respectively. Thus, there can be up to undefined number of centers in reality.

(b) If, in the opinion of the authors, Trp-172, Trp-215 and Trp-141 are efficiently deactivated by non-chromophore groups, then we cannot consider these tryptophan residues as equally participating in the migration within a group: their intrinsic and sensitized fluorescence should be very low, hence, these residues cannot be efficient donors in energy migration. Moreover, the authors used in the Forster equations (11.5 and 11.6) for all tryptophan residues the same value of fluorescence quantum yield, 0.14, that is wrong.

(c)   The excitation wavelength was 296 nm in the experiments. However, it is well known that red excitation results in the Weber's effect, i.e., a sharp break in the migration [7, 60, 289]. It is generally accepted that no energy migration between tryptophans occurs when proteins are excited at wavelengths above 295 nm [17, 313]. Thus, one may conclude that either the authors carried out the experiments under the conditions where no migration could take place or the generally accepted point of view is incorrect.

(d)   The distances between tryptophans within groups were no more than 10 Å whereas the size of tryptophan chromophore is about 7 Å. Thus, the dipole-dipole approximations is inapplicable.

(e)   Although the authors mentioned that the lack of information about individual spectral properties of each tryptophan led to some uncertainty in calculations, nevertheless, they arbitrarily assumed the overlap integral to be the same for all the tryptophans.

(f)   The overlap between the tryptophan absorption and emission spectra is negligible. This indicates that cold migration is actually impossible (Chapter 11): vibrational relaxation significantly decreases the energy of the singlet excited state of tryptophan. Therefore, the energy of the majority of emitting centers is substantially lower than the energy levels of the absorption centers. In general, hot energy migration between closely spaced tryptophan residues is possible. However, there is no reliable experimental evidence in favour of this assumption.

Attempts have been made to recognize energy migration along tryptophans using fluorescence depolarization and quantum yield [15–18, 313]. Results obtained in these studies are contradictory and ambiguous for the following reasons:

(a)   Tryptophan and tyrosine residues are both excited with the wavelength of 280 nm. Tyrosine and tryptophan residues in one macromolecule can be hypochromised. Quantitative estimation of hypochromism in proteins is a very difficult task (Chapters 2 and 3).

(b)   No migration can occur with the red-edge excitation of tryptophan at 295 nm (Weber's effect).

(c)   Some amino acid residues play an important role quenching tryptophan fluorescence in proteins.

(d)   It has been found that the main reason of depolarization of tryptophan fluorescence in proteins is the rotational mobility of tryptophans, protein segments, and protein molecule as a whole.

(e)   Simultaneously with the equilibrium thermal mobility, after photoexcitation of tryptophans, an induced mobility appears. This increases depolarization.

13.3 Tryptophan–NADH Pair in Alcohol Dehydrogenase

Let us consider the quenching of fluorescence of horse liver alcohol dehydrogenase while binding to NADH.

Previous studies [147] show that quenching of tryptophan fluorescence in alcohol dehydrogenase while binding to NADH is accompanied by energy transfer to the pyridine ring of NADH. The efficiency of energy transfer from Trp-314, localized nearby pyridine ring of NADH, and from Trp-15, 27 Å apart from the pyridine ring of NADH, was not precisely estimated for a long time. Furthermore, it has been found that the quenching of fluorescence of alcohol dehydrogenase

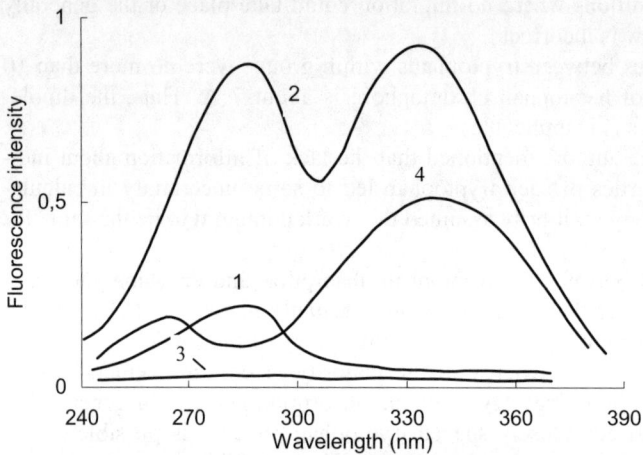

Fig.13.2. Excitation spectra of alcohol dehydrogenase (1), of NADH bound to alcohol dehydrogenase (2), of NAD^+ in solution (3), and of NADH in solution (4). Concentration of alcohol dehydrogenase is 0.14 mg/ml, of NADH is 4 µM, of NAD^+ is 20 µM. The emission is registered at 435 nm [165]

occurs also after adding NAD^+ [320–322]. NAD^+ has no absorption band at 340 nm, and, therefore, no Forster energy transfer to NAD^+ is possible. Some authors assume that NAD^+ induces considerable conformational changes in alcohol dehydrogenase, that results in quenching. We earned out an extensive study to estimate the efficiency of energy transfer and to understand the role of conformational changes in quenching while binding to NADH or NAD^+ [60, 165, 323].

Fig. 13.2 shows the excitation spectrum of NADH bound to the enzyme when quenching is maximal, about 50%. When protein spectrum and NADH are substracted from the total spectrum, we receive the intensive band at 286 nm This band at 286 nm reflects the energy transfer from tryptophan residue (the corrected excitation spectra correspond to the absorption spectra; the extinction coefficient of tryptophan in the protein is similar to those of DL-tryptophan).

The maximum value of quenching of alcohol dehydrogenase after binding to NADH depends on the wavelength of excitation and emission [165, 321, 324]. The dependence of quenching on the excitation wavelength could be explained by the difference in absorption (excitation) spectrum of Trp-314 from that of Trp-15.

Probably, it is due to the Trp-314 hypochromism [165], provoked by the close proximity and parallelism of the two Trp-314 residues of subunits of the enzyme. The quenching depends on the emission wavelength. Trp-314 has its maximum at

13.3 Tryptophan–NADH Pair in Alcohol Dehydrogenase

324 nm and is quenched, whereas Trp-15 has its maximum at 335 nm and is not quenched.

The theoretical value of the Forster radius for the Trp-314-NADH and Trp-15-NADH pairs can be calculated using (11.5), (11.6). Table 13.2 represents the values of R determined by X-ray crystallography [153], the calculated R_0 values and the experimental and expected theoretical values of the energy transfer efficiency (E_e and E_t) [165, 323]. The values of E_t were calculated using the Forster approach; the values of E_e were determined using emission and excitation spectra by the decrease in the quantum yield of the donor and the increase in the luminescence intensity of the acceptor.

Table 13.2. Distances between chromophores in alcohol dehydrogenase (as determined by X-ray diffraction), Forster radii, experimental and theoretical values of energy transfer efficiency

Potential donor	Potential acceptor	R, [Å]	R_0, [Å]	E_t, [%]	E_e, [%]
Trp-314(a)	Trp-314(b)	6	11	97.5	0
Trp-314(a)	Trp-15(a)	41	11	0	-
Trp-314(a)	Zn(a)	29	9	0.7	-
Trp-314(a)	Pyridine in NADH(a)	17	36	98.9	100
Trp-15(a)	Pyridine in NADH(a)	27	32	73.5	0
Trp-314(a)	Pyridine in NADH(b)	22	36	95.1	0

(a) and (b) are subunits of the enzyme

When the values of R_0 are large (36 and 32 Å) the energy transfer is supposed to be highly efficient. However, an efficient energy transfer was observed only in

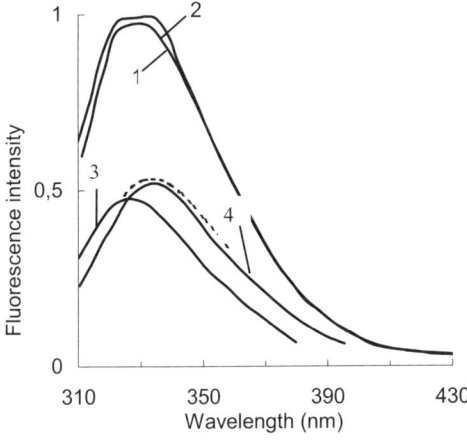

Fig. 13.3. Emission spectra of alcohol dehydrogenase excited at 295 nm (1) and at 286 nm (2), the band of Trp-314 (3) and Trp-15 (4), and the emission spectrum of alcohol dehydrogenase after quenching by NADH (-----) [323]

the pair Trp-314(*a*)-NADH(*a*), where the chromophores were very close to each other. In the pair Trp-15(*a*)-NADH(*a*) no energy transfer was observed. The results are inconsistent with the inductive resonance Forster calculations, according to which the efficient energy transfer is expected to occur in the pair Trp-15(*a*)-NADH(*a*) and in the pair Trp-314(*a*)-NADH(*b*). Also, no any energy transfer from tryptophan residues to NADH was found in rabbit muscle aldolase [218].

Fig. 13.3 shows the bands of Trp-314 and Trp-15 in the total fluorescence of alcohol dehydrogenase as well as the spectrum recorded after adding NADH. Decomposition of the emission spectrum of alcohol dehydrogenase in two components is described in [165, 321, 324]. The intensities of the components obtained in [165] differ from those in [321, 324] because the ratio between Trp 314 and Trp-15 emission depends on the excitation wavelength [165, 321, 323, 324].

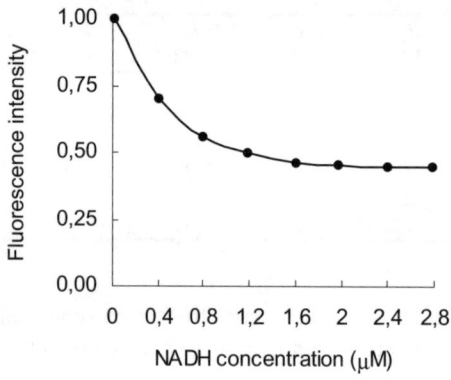

Fig. 13.4. Dependence of the quenching of alcohol dehydrogenase fluorescence on the concentration of added NADH. Excitation is at 286 nm, emission is at 328 nm, monochromator slits are 4 nm each. The concentration of enzyme is 0.035 mg/ml [165]

Fig. 13.4 demonstrates the dependence of quenching of alcohol dehydrogenase on the concentration of NADH. Maximum quenching is obtained when two NADH molecules are bound to one molecule of the enzyme (two subunits), and this value does not change with further increase in the concentration of NADH. Trp-314 but not Trp-15 is quenched.

The conclusion that only Trp-314 is quenched by NADH was supported by phase-modulation measurements. The lifetime at the wavelength of 320 nm decreased from 4.0 ns to 2.8 ns and at 360 nm from 4.8 ns to only 4.5 ns under of maximum quenching. If the ratio between Trp-314 and Trp-15 amplitudes are taken into account, we obtain that of Trp-15 is not quenched.

Since the value of maximum quenching of alcohol dehydrogenase fluorescence after binding to NADH is almost entirely determined by the energy transfer, we may conclude that the conformational changes play no significant role in quenching.

The values of maximum quenching of the enzyme fluorescence by some coenzymes, substrates, and their analogs are summarized in Table 13.3. Only NADH is a strong quencher, but not other compounds. On the basis of the obtained data we conclude that no substantial conformational rearrangements in enzymes are observed.

A question arises. What is the reason of quenching observed after adding NAD^+?

Fig. 13.5 demonstrates the changes in the spectrum of non-dialyzed alcohol dehydrogenase after the addition of 20 µM NAD^+ and a subsequent addition of acetaldehyde. The band at 435 nm, which appears after addition of NAD^+, does not differ from the corresponding band in the spectrum of the complex between the

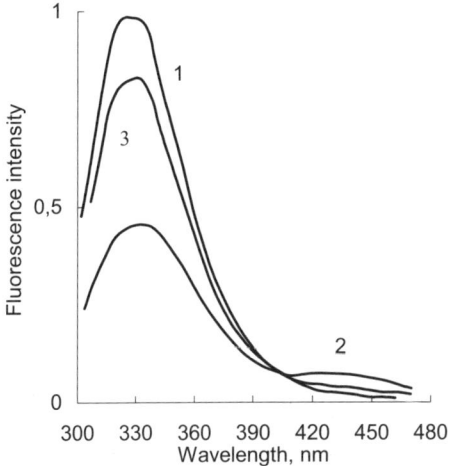

Fig. 13.5. Changes in the fluorescence spectrum of alcohol dehydrogenase after addition of NAD^+ and acetaldehyde [165]: 1) 0.14 mg/ml of the enzyme (without dialysis), 2) 20 µM NAD^+ were added to the enzyme, 3) 500 µM acetaldehyde were added to the enzyme after adding 20 µM of NAD^+. Excitation is at 286 nm

enzyme and NADH. The excitation spectrum registered at 435 nm is identical to spectrum 2 in Fig. 13.2. If alcohol dehydrogenase was dialyzed, NAD^+ quenched its fluorescence by no more than 15% (Table 13.3). Probably, when NAD^+ is bound to non-dialyzed alcohol dehydrogenase, it is reduced to NADH, which quenches the fluorescence of tryptophan in the enzyme by energy transfer. Acetaldehyde, which oxidizes NADH in the enzyme, transforms NADH to NAD^+, so that the quenching is almost entirely eliminated (Fig. 13.5). Traces of ethanol used in the enzyme preparation probably serve as a reducing agent for NAD^+ bound to the protein. It should be noted that the preparation of NAD^+ was not contaminated by NADH (as determined by spectrophotometric and fluorescent techniques).

The 15% quenching of alcohol dehydrogenase fluorescence by NAD^+ may be explained by (a) formation of an exciplex between Trp-314 and NAD^+ when their

Table 13.3. Values of maximum quenching of alcohol dehydrogenase fluorescence by various substrates [323]

Substrate	Concentration of substrate at maximum quenching, [μM]	Maximum quenching, [%]
NADH	4	55
NAD^+	20	15
Ethanol	1000	0
ADP	500	5
Adenosine	50	0
Acetaldehyde	500	5

Conditions: Concentration of alcohol dehydrogenase is 0.14 mg/ml. Excitation is at 286 nm, emission is at 329 run

chromophores are in contact, (b) deactivation of Trp-314 by NAD^+, when they are in contact. In favour of the direct contact is the evidence obtained by Schellenberg [325] indicating the incorporation of deuterium from the substrate in the tryptophan of yeast alcohol dehydrogenase. Moreover, it was shown [326] that NAD^+ is the effective «contact» quencher of phosphorescence in horse liver alcohol dehydrogenase. Also, it should be kept in mind that the photoexcitation of tryptophan in the enzyme induces an additional mobility of the indole chromophore and its surrounding (Chapter 8), so that the distance between Trp-314 and the pyridine ring of NAD^+ or NADH can be closed on a nanosecond time scale.

In our fluorescence experiments, no substantial conformational changes were observed in the enzyme after binding NADH, NAD^+ or other substrates to the enzyme. This result is in good agreement with the results of high-resolution X-ray diffraction analysis for the NADH-enzyme complex [327], indicating that the binding of NADH to alcohol dehydrogenase induces only insignificant structural changes in polypeptide residues; small changes were observed in the region between N 293 and N 297 and no changes were found in other regions of the enzyme (no precise X-ray diffraction data on the enzyme-NADH complex were reported before the obtained spectroscopic results on alcohol dehydrogenase were published [165, 323]).

13.4 Tryptophan–Heme Pair in Myoglobin

Heme proteins, particularly myoglobins, are structurally the most "rigid" proteins. The tryptophan fluorescence in hemoglobin and myoglobin is strongly quenched by the heme [533]. This quenching is generally attributed to the transfer of electronic excitation since the spectrum of tryptophan fluorescence overlaps with the UV absorption spectrum of the heme at 330 nm.

However, the quenching itself is not a sufficient argument in favor of the electronic excitation energy transfer. Indeed, according to the golden rule (11.3) the transfer constant depends on the frequency. The longer is the wavelength, the higher is the transfer rate. Therefore, the deactivation process (due to multiple energy interactions with IR frequencies), where the electronic excitation energy spend in many vibrational modes of the acceptor is more feasible than electronic energy transfer. It can be assumed that the deactivation by heme in proteins competes with resonance energy transfer to heme.

The average lifetime of the tryptophan fluorescence in sperm whale myoglobin does not exceed 0.5 ns (Table 13.4) [181]. This makes the rate of quenching no less than 2×10^9 s^{-1}. The quantum yield of the tryptophan fluorescence in myoglobin is approximately 20 times lower than that in apomyoglobin, 0.005–0.007 and 0.12 [181], respectively; this is consistent with the results obtained by Andersen [511]. These results indicate "instantaneous" quenching with the rate constant ~ 10^{10}–10^{11} s^{-1}. In horse heart myoglobin at neutral pH [533] three main lifetime components in the Trp emission were found: 0.04, 0.11, and 1.3 ns (plus a minor

Table 13.4. Lifetimes of fluorescence of myoglobin, apomyoglobin and apomyoglobin-protoporphyrin IX complex

Sample	Lifetime, [ns]	
	at 330 nm	at 628 nm
Myoglobin	≤0.5	<<0.1
Apomyoglobin	2.5	–
Apomyoglobin-protoporphyrin complex	<2.5	18.0
Protoporphyrin in water	–	12.0

Excitation wavelength is 286 nm for tryptophan and 409 nm for porphyrin

component of 4.8 ns). Therefore, the emission must be independent on the protein dynamics. Indeed, the fluorescence intensity of myoglobin does not depend on temperature (Table 13.5).

Heme does not fluoresce. Thus, it makes us impossible to detect resonance energy transfer by sensitized fluorescence of the acceptor and to differentiate resonance energy transfer from other quenching processes. The complexes of apohemoglobin [328] and apomyoglobin [181] with protoporphyrin IX were prepared. In these complexes the sensitized emission of the acceptor could be measured.

The authors [328] said that the excitation spectrum of the porphyrin emission of the complex was directly proportional to the absorption spectrum in the visible and UV region within an experimental error estimated to 5%. They concluded that this result indicates the 100% energy transfer efficiency from Trp and Tyr to porphyrin. Unfortunately, no the excitation spectrum was shown in the paper.

The excitation spectrum of protoporphyrin IX in complex with apomyoglobin was almost alike to the absorption spectrum of the complex [60]. However, the intensity of the donor band of tryptophan at 280 nm was much lower than one would expect assuming 100% efficiency of electronic excitation energy transfer. In fact, the efficiency was found to be 50% of the quenching value. Probably, simple deactivation, i.e. an exchange of electronic excitation for many vibrational

Table 13.5. The intensity of tryptophan fluorescence in sperm whale myoglobin at different temperatures

Temperature, [°C]	Fluorescence intensity, [%]
8	100
14	98
25	98
36	98
47	98
58	98

quanta in the acceptor, protoporphyrin IX, successfully competes with the process of electronic energy transfer. This may indicate efficient direct deactivation where two tryptophans at the N-end of the protein are separated from the protoporphyrin or heme by a distance of about 20 Å.

13.5 Tryptophan–Pyrene Pair

The tryptophan fluorescence spectrum strongly overlaps with the absorption spectrum of pyrene, which has an extinction coefficient of 56 000 M^{-1}cm^{-1} at 334 nm. Thus, the Forster radius, R_0, equals to 28.6 Å [137]. Binding of pyrene with proteins results in quenching of the tryptophan fluorescence. However, the efficiency

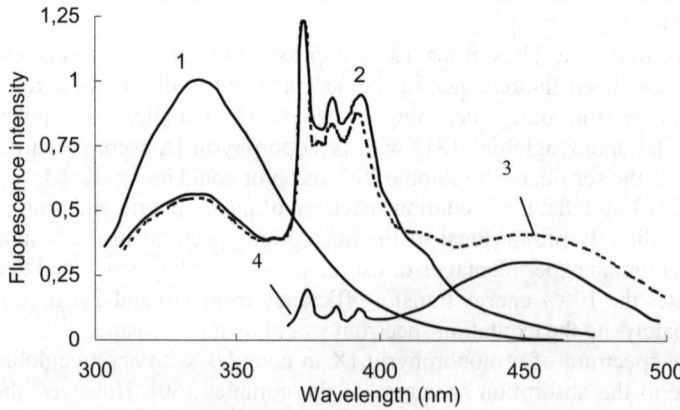

Fig. 13.6. Emission spectra of albumin (1), albumin with pyrene (2, 3) and pyrene without the protein (4). Concentration of albumin is 0.2 mg/ml, concentration of pyrene is 3.6 µM (2) and 7.2 µM (3, 4). Excitation wavelength is 286 nm; 0.4-cm cuvette was used [329]

of resonance energy transfer is low as compared with the efficiency of quenching. The quenching of tryptophan fluorescence in proteins by pyrene is most likely to proceed by deactivation or exciplex formation.

Bovine serum albumin has two tryptophan residues [15–17], one of which is buried inside the globule, and the other is located near the surface and easily accessible for various quenchers [15–18, 60]. Binding of pyrene results in complete quenching of this accessible tryptophan. The buried tryptophan is not quenched at all no matter how large is the concentration of pyrene (Fig. 13.6 and Fig. 13.7).

Probably, the quenching occurs when pyrene comes in contact with the accessible tryptophan.

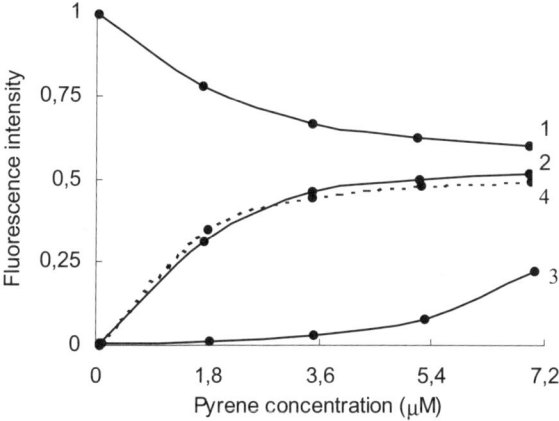

Fig. 13.7. Effect of pyrene concentration on the fluorescence intensity of bovine serum albumin and pyrene. 1) protein fluorescence at 340 nm; excitation at 286 nm; 2) monomer pyrene fluorescence at 393 nm; excitation at 286 nm; 3) dimer pyrene fluorescence at 470 nm; excitation at 286 nm; 4) monomer pyrene fluorescence at 393 nm; excitation at 334 nm. Concentration of albumin is 0.2 mg/ml [329]

Fig. 13.6 shows the fluorescence spectra of albumin and pyrene before and after binding. The monomer band at 380-400 nm and the band of the pyrene dimer at 400-500 nm are seen. The increase of the monomer fluorescence of pyrene resulting from binding to albumin is not related to energy transfer. This increase is due to the increased lifetime and quantum yield of pyrene in the protein hydrophobic microenvironment as compared with pyrene in the polar aqueous phase.

The maximum efficiency of quenching of albumin fluorescence by pyrene reaches about 44% (Fig. 13.7). Each protein or enzyme has a maximum value of quenching efficiency. This value is a constant that characterizes the steric accessibility of indole chromophores to the quencher. For pyrene, the maximum efficiency is achieved when the molar ratio of pyrene to protein is no less than 1:1. Simultaneously with the decrease of the intensity of protein fluorescence, the lifetime becomes smaller. For bovine serum albumin, the lifetime decreased from 5.4

to 3.8 ns.

A significant decrease in the lifetime when tryptophan fluorescence is quenched coupled with the fact that the efficiency of electronic excitation energy transfer is small, points to the deactivating mechanism of quenching in albumin and probably in other proteins. Thus, the quenching of tryptophan fluorescence in proteins by pyrene is most likely to proceed by deactivation or exciplex formation.

Fig. 13.8 shows the excitation spectrum of pyrene bound to bovine serum albumin and to lecithin liposomes in the range between 240 and 320 nm [329]. The band at 285 nm in the spectrum of pyrene bound to albumin is appeared due: (a) resonance energy transfer; (b) the emission of the directly excited pyrene; (c) reabsorption of the protein fluorescence by pyrene; (d) volume reabsorption; (e) flu-

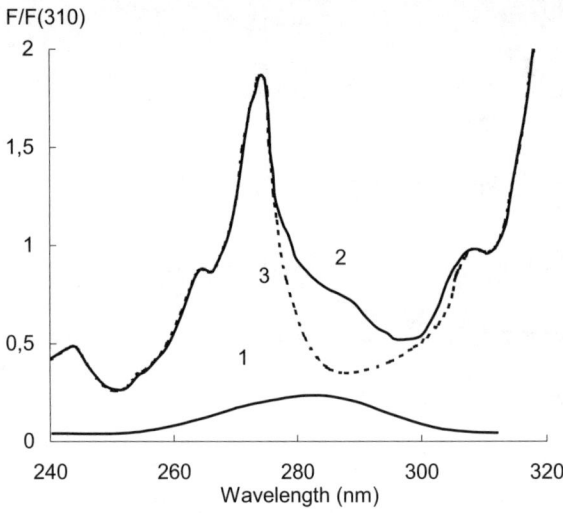

Fig. 13.8. Excitation spectra (un-corrected) of albumin (1), pyrene bound to albumin (2), and pyrene bound to liposomes (3). Concentration of albumin is 0.5 mg/ml, concentration of liposomes is 0.2 mg/ml, concentration of pyrene is 7.2 µM. The wavelength of emission is 393 nm for all spectra [329]

orescence of the protein at 393 nm; (f) emission of pyrene dimers having absorption bands at ~284 nm and 360 nm (Fig. 13.9). Quantitative estimation of these effects indicates that the efficiency of resonance energy transfer is low.

Quantitative estimation of the intensity of the band at 286 nm in the excitation spectrum of pyrene bound to albumin based on the absorption and excitation spectra and the value of quenching efficiency, indicates that if the quenching was due to the energy transfer from tryptophan residues to pyrene, the intensity at 285 nm would be ~2 times greater than that observed experimentally. Very low efficiency of energy transfer (~1%) from the tryptophan residues in apolipoprotein B-100 to the pyrenyl-phospholipid probes was found in [330].

When the molar ratio of pyrene to albumin is greater than 1:1, pyrene molecu-

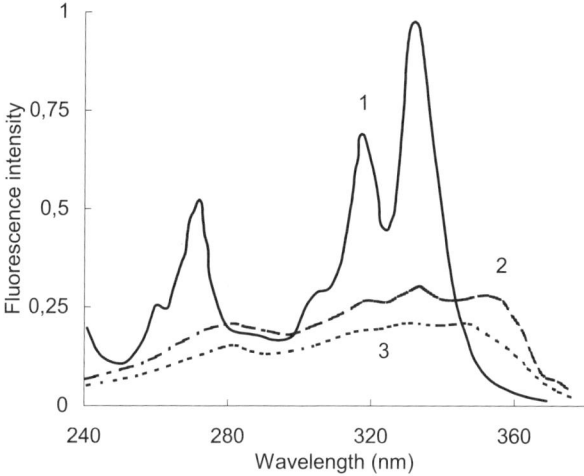

Fig. 13.9. Excitation spectra of pyrene monomers and dimers [60, 329]: 1) pyrene with 0 1 mg/ml bovine albumin, emission at 393 nm, 2) pyrene with 0 1 mg/ml albumin, emission at 470 nm; 3) pyrene without protein, emission at 470 nm Concentration of pyrene is 18 µM

les are dimerized. Dimerization is detected by the appearance of structureless emission at 460 nm (Fig. 13.6). The higher is the concentration of dimers, the slower is the increase in the luminescence intensity of the pyrene monomers. The excitation spectrum of the dimer significantly differs from that of the monomer (Fig. 13.9). All bands in the dimer spectrum are broadened and hypochromised. The band at 360 nm was used in [331] for detection of a minor content of the pyrene dimers in polymers. The band at 285 nm in the excitation spectrum does not belong to tryptophan, since this band is observed while pyrene dimerizing in aqueous solution without protein. Dimerization mainly takes place in water or perhaps in the polar regions of the protein rather than in the hydrophobic pocket where one molecule of pyrene is bound. Binding of pyrene in the hydrophobic pocket of albumin protects pyrene from being quenching by oxygen; the lifetime of pyrene bound to albumin is 225 ns.

13.6 Quenching of Tryptophan Emission by Dyes

The efficiency of energy transfer strongly depends on the interchromophore distance. This is used to estimate distances in biological systems, in particular, in proteins and multi-enzyme complexes. Fluorescent labels or probes bound to proteins are often used for this purpose [332–335]. Various organic synthetic dyes, sometimes aromatic hydrocarbons and rare-earth ions, serve as labels and probes. In the majority of studies of energy transfer authors consider it appropriate to apply Forster theory, though, as a rule, no special analysis of the transfer mechanism is conducted. Quenching of protein fluorescence by chromophore compounds such as probes, dyes, aromatic hydrocarbons is usually considered as the result of inductive resonance energy transfer [67, 19, 21, 23, 24].

For instance, in [336] a single cysteine residue located in the hydrophobic membrane anchor domain in colicin El COOH-terminal channel peptide was labeled with the thiol-specific fluorescent reagent IAEDANS. The eleven single Trp mutants of the channel peptide were prepared, and the energy transfer for each Trp residue to the AEDANS chromophore was measured (R_0=20–26 Å). The transfer efficiency values were ranging from the excitation spectra using Forster's theory. The transfer efficiencies ranged from 15% to 100%, that corresponded to the distances between 8 and 46 Å. Activation of the channel peptide to the insertion-competent state by adding octyl β-D-glucoside resulted in decreased transfer efficiencies. A direct correlation was observed between the change in the transfer efficiency caused by the channel peptide activation and the position of the Trp residue within the channel peptide primary sequence.

The following remarks could be made on the results of this work. As one can see from the spectra in [336], the difference in the excitation spectra at 280 nm is not only due to energy transfer. Moreover, the spectrum of AEDANS (without Trp) is not shown in the paper. The authors state that since no Trp emission is detected in W-507-AEDANS complex, the energy transfer efficiency is equal to 100%. In contradiction to this statement, one can see from spectra in [336] that the intensity of Trp fluorescence in W-507-AEDANS is not zero, and it is equal to about 15% of W-507 fluorescence. The authors say that the second method, i.e., the measurement of donor lifetimes, gave similar results. Unfortunately, no lifetime data were represented in the paper.

The tryptophan fluorescence of human apo-cyclooxygenase-2 was quenched (on 40%-50%) by the inhibitors: diclofenac, indomethacin, ketoprofen, and etc. [337]. It was suggested that the cause of quenching is resonance energy transfer to the inhibitors. Using Forster theory, the distances were calculated. However, no special study of quenching mechanisms were made. Quenching was identified with the efficiency of resonance energy transfer without data on sensitized fluorescence of the inhibitors.

Phosphorescence and fluorescence energy transfer from the tryptophan residue of the mitochondrial soluble part of H^+-ATPase complex (F_1) to the two probes were detected in [338]. The first fluorescence probe, 2'-o-(trinitrophenyl)adenosine-5'-triphosphate, was bound to the nucleotide binding sites of the enzyme, whereas the second probe, 7-diethylamino-3'-(4'-maleimidylphenyl)-4-methylcoumarin, was attached to the single sulfhydryl residue of isolated oligomicin sensitively-conferring protein, which was then reconstituted with F_1. The efficiency of energy transfer was estimated not only by the Trp quenching but also by sensitized emission of the probes.

The following general comments on many of energy transfer measurements should be pointed out. First, efficient energy transfer usually takes place when the donor is separated from the acceptor by a small distance, about 10 Å or less [162]. However, it is known that at small donor-acceptor distances one of the most efficient ways of transformation of electronic excitation is the formation of an exciplex [57, 162]. Exciplex formation may successfully compete with the transfer of electronic excitation. Second, with a few exceptions, the quantum yield of the exciplex emission is low. Therefore, it is not always possible to register the exciplex emission because of the intense fluorescence of protein, dye or aromatic hydrocarbon. Third, along with the processes of electronic excitation transfer and fractional energy transfer in exciplexes, a complete deactivation of the excited donor (at dynamic or static quenching) can take place [339].

There are a number of typical mistakes often made in biochemical studies on energy transfer using fluorescent probes. They are [60, 162]: (a) Energy transfer is traditionally described in terms of the Forster theory, although no special study of quenching mechanisms were made in the majority of works. (b) Quenching is identified with the efficiency of resonance energy transfer without quantitative data on sensitized fluorescence of the acceptor. (c) Probes and labels have the si-

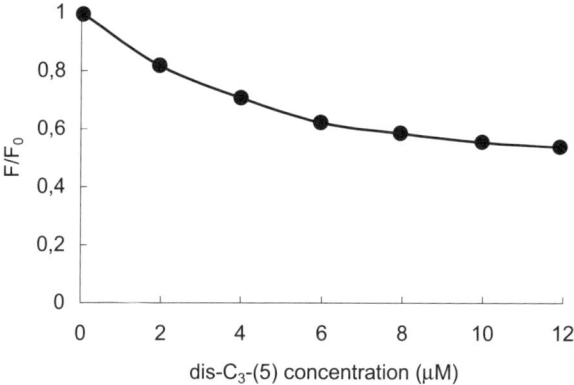

Fig. 13.10. Normalized integrated fluorescence intensity of BSA tryptophan as a function of the concentration of added diS-C_3-(5) [339]

zes comparable to the distances (i.e., the dipole-dipole approximation is not suitable). (d) The orientation factor is arbitrarily taken to be 2/3.

Quenching of fluorescence from tryptophan in proteins by dyes and other chromophoric compounds is usually explained as the result of energy transfer. However the overlap between the Trp emission spectrum and the absorption spectrum of a dye is one of the conditions required for resonance energy transfer (and hence for the extent of quenching) but this is not sufficient to explain all aspects of quenching. Competitive processes (dynamic deactivation, exciplex formation and others) are in many cases dominant, as was shown for potential Forster donor-acceptor pairs: Tyr-Trp, Trp-pyrene and Trp-NADH.

In order to illustrate these effects, bovine serum albumin (BSA) was selected because of its extraordinary ligand-binding properties. BSA has two tryptophan residues: Trp 135 and Trp 214. Trp 135 is buried inside the globule, while Trp 214 is located near the surface and hence more accessible for various quenchers. Almost all dyes used at concentrations of ~1–10 µM, quench the Trp fluorescence of BSA (Table 13.6) [339]. Typical dissociation constants (Kd) are therefore of the order of 10^{-6} M, which is not unusual. No increase in Tip fluorescence was observed in the experiments. Any spectral shift of the remaining Trp emission was absent. Therefore it was concluded that the binding of dye molecules does not cause a global change in the protein conformation.

Negatively charged and uncharged dye molecules result in more quenching of Trp fluorescence in albumin than positively charged dyes. This observation is due to the presence of one (or more) positively charged group(s) in the binding pocket, at which both Trp residues are located in the near-vicinity. The three-dimensional structure is composed of three homologous domains (I, II, and III). Each domain is subdivided in to two helical subdomains (A and B) which are connected by disulfide bridges. The site where negatively charged aromatic residues are bound is localized in a hydrophobic pocket in subdomain IIA. Subdomain IIA indeed contains a positively charged group, namely, Lys-199.

The limiting value of quenching, reached at a molar excess of a dye with respect to protein (Fig.13.10), is characteristic for each dye. The limiting value is constant and reflects the accessibility of Trp for a given dye. A high quenching activity was detected (in decreasing order): eosin, thiazine red, ANS, titanium yellow, trypan red, diS-C_3-(5), pyrene, and thymol blue. Among these, diS-C_3-(5) has no light absorption in the region of the Trp emission: the spectral overlap integral is about zero. Parallel to a decrease in the Trp gfluorescence in BSA during ti-

Table 13.6 Data from Quenching Experiments of BSA Tryptophan Fluorescence by Different Dyes [339] [a]

Dye	Charge	RSOI [%]	1–(F/F_0) [%]	F_{dye} [rel. units]
Eosin	2-	30	80	2
Thiazine red	2-	47	76	0
ANS	1-	20	70	1
Titanium yellow	2-	100	64	1
Trypan red	5-	19	48	0.2
DiS-C_3-(5)	1+	~0	46	0.5
Pyrene	0	25	44	1
Thymol blue	1-	28	32	0
Acridine orange	1+	~0	24	1
Pyronine B	1+	19	15	0.1
Chryzoidin	1+	75	14	0
Coryphosphine	1+	38	6	3
Auramine 00	1+	9	0	0.05

[a] The charge of the dye molecules is given, together with the relative spectral overlap integral (RSOI) of Trp emission and dye absorption, the amount of quenching 1–(F/F_0), and the relative dye fluorescence upon excitation at 286 nm. The overlap integral $\int F(v)E(v)dv$ is contained in the critical transfer distance R_0. The overlap integral for the pair Trp-titanium yellow is largest and, for that reason has been set to 100%

tration with dye molecules, a reduction of fluorescence lifetimes was observed. For instance, saturating binding of pyrene or diS-C_3-(5) reduces the average fluorescence decay time of Trp from 5.5 to 3.8 ns. Quenching of Trp fluorescence in BSA by aromatic hydrocarbones and dyes most likely proceeds via radiationless deactivation and exciplex formation.

Binding of eosin, ANS, titanium yellow, acridine orange, coryphosphine, and pyrene to BSA resulted in a sharp increase in their fluorescence (Table 13.6). This emission was not sensitized (to a considerable extent) but was caused by direct excitation of the dye molecules at 286 nm. Trypan red, pyronine B, auramine 00, and diS-C_3-(5), upon binding to BSA, show a weak fluorescence upon 286-nm excitation. The latter fluorescence intensity was much lower than the decrease in Trp fluorescence, indicating that the quenching is caused by both radiationless deactivation and resonance energy transfer. In the case of binding of thiasine red, thymol blue, and chryzoidin to BSA, the fluorescence of these dyes at 286-nm excitation was not observed, although quenching of Trp fluorescence took place. The latter observation indicates that the quenching is caused preferentially by radiationless deactivation via both chromophoric and nonchromophoric groups in these dye molecules.

13.7 Conclusion

Efficient energy transfer in proteins occurs over short distances comparable with the sizes of the chromophores. The efficiency of the transfer of electronic excitation energy is usually lower than the efficiency of quenching due to competing processes. Quenching of the donor fluorescence often results from donor deactivation via dynamic or static mechanisms. A correlation between the limiting quenching values and the Forster spectral overlap integral was not found. Forster theory does not provide an adequate quantitative description of energy transfer in the above systems. For instance, the calculated energy transfer from Trp-15 to NADH separated in alcohol dehydrogenase by ~27 Å was not observed. Also no considerable energy transfer from tyrosine to tryptophan was detected in proteins.

14 Energy Transfer in Biomembranes

Energy transfer between fluorescent probes or labeles is widely used to study a spatial organization and function of biomembranes [338, 340, 341–344]. Review and analysis of the literature on energy transfer between fluorescent dyes was made in [60, 162].

To measure the efficiency of energy transfer in membranes is a very difficult task. Quenching usually is ascribed to energy transfer. For instance, in [345] energy transfer between labeled glycophorin-A peptides was used to observe their dimerization in lipid bilayers. Peptides were labeled with 2,6-dansyl chloride as the donor and dansyl chloride as the acceptor. The efficiency of energy transfer was identified by the authors with the quenching of fluorescence intensity. Such identification without estimation of sensitized emission is not correct.

The efficiency of energy transfer from three tryptophan residues to the N'-(iodoacetyl)-N'-(5-sulfo-1-naphthyl)ethylenediamine label in five mutants of colicin-A inserted in lipid vesicles was measured by a sensitized fluorescence of the label [340]. The labeling was not 100 %, but there was not any other information about it in the article. The distances were calculated according to Forster model. Unfortunately, the control data without the label, without the donors, and the emission spectra were not shown.

Tryptophan fluorescence of adrenocorticotropin was quenched by 3-(9-anthroyloxyl) stearic acid in phospholipid vesicles [312]. This quenching was interpreted as a result of resonance energy transfer. However, no data on the anthroyloxyl fluorescence were represented in the paper. The quenching could be a collisional nature. It is known that the excited anthroyloxy-labeled fatty acids can move across the lipid bilayer [346].

Using Forster theory, a phenomenological model of energy migration between identical fluorophores in membrane protein oligomers was developed [344]. The model was tested on the fluorescein-labeled mellitin oligomers. The experimental values of anisotropy were close to the calculated values when non-emitting dimers were taken into account. Unfortunately, lifetime measurements were not carried out. These measurements obligatory need, since decrease in anisotropy could occur not only during the energy migration but also due to a lifetime increase upon oligomerization.

Frequency-domain fluorescence energy transfer measurements between pyrene maleimide located at Cys-27 in calcium binding loop I and nitrotyrosine-139 in calcium binding loop IV on wheat germ calmodulin indicated that the average spatial separation and conformational heterogeneity associated with the two opposing globular domains of calmodulin are virtually identical upon calmodulin binding to either the plasma membrane Ca^{2+}-ATPase in native erythrocyte ghost membranes or a peptide that has an identical sequence [347]. The calculations were curried out in frameworks of the Forster theory. Nitrotyrosine is a bad emitter; therefore, its sensitization was not measured. The R_0 value for the calmodulin - erythrocyte ghosts system was of 14.7 Å. The distance of 20.2 Å was found. These values are only in 2–3 times more than the size of the used chromophores, i.e. the dipole-dipole Forster approximation is hardly suitable.

14.1 Quenching of Tryptophan Fluorescence in Sarcoplasmic Reticulum by Probes

One of the first applications of quenching in the study of the structure of membrane proteins was described in [348]. Quaternary salts of 4-picoline served as the efficient collisional quenchers of the tryptophan fluorescence of membrane proteins, only surface tryptophan residues are quenched by these compounds. Fluorescence quenching in sarcoplasmic reticulum and erythrocyte membranes was measured. In sarcoplasmic reticulum the fraction of accessible Trp residues was 54 %. The fluorescence component which was quenched by N-methylpicolinium perchlorate was taken as an index of the membrane protein exposure to the aqueous environment. Later a similar approach was applied in the study of the structural organization of membrane proteins in sarcoplasmic reticulum, microsomes and mitochondria using pyrene [137], pyrene-butyrate, 4-dimethylaminochalcone,

Table 14.1. Values of maximum quenching of tryptophan fluorescence by various probes in membranes of longitudinal tubules and terminal cisternae of sarcoplasmic reticulum [60]

Qencher	Concentr. [μM]	Charge	Maximum quenching [%]		Energy transfer efficiency [%]
			LT	TC	
Cesium	50000	1+	13	9	-
Calcium	500	2+	0	0	-
Pyrene	7	0	65	54	~0
Anthracene	5	0	29	21	~0
ANS	10	1-	48	35	≤20
Titan yellow	1	3-	29	28	<10
Trypan red	1	5-	32	29	<10
Auramine 00	4	1+	8	7	0
Pyronin B	4	1+	9	9	0
Coriphosphine	5	1+	27	18	<10
di-S-C$_3$-(5)	10	1+	52	42	~0
Caffeine	500	0	0	0	-

Conditions: Excitation is at 286 nm, emission is at 340 nm. Protein concentration for membranes in solution is about 1 μM. Medium: 5 mM imidazole, 100 mM sodium chloride, 2.5 mM magnesium chloride, pH 7.0. The data were corrected for screening and reabsorption

ANS, and 2,4-dinitrophenol [60] as quenchers.

Fluorescence quenching of tryptophan residues of Ca^{2+}-ATPase from muscle sarcoplasmic reticulum by cholesterol-containing phosphatidylcholines [516], dyes and aromatic hydrocarbons [60] shows that the residues probably bind on the enzyme-lipid interface.

Magnitudes of fluorescence quenching of tryptophan by various dyes, aromatic hydrocarbons and cations for two fractions of sarcoplasmic reticulum, i.e., longitudinal tubules (LT) and terminal cisternae (TC), are summarized in Table 14.1. The value of quenching depends on the physico-chemical properties of the mem-

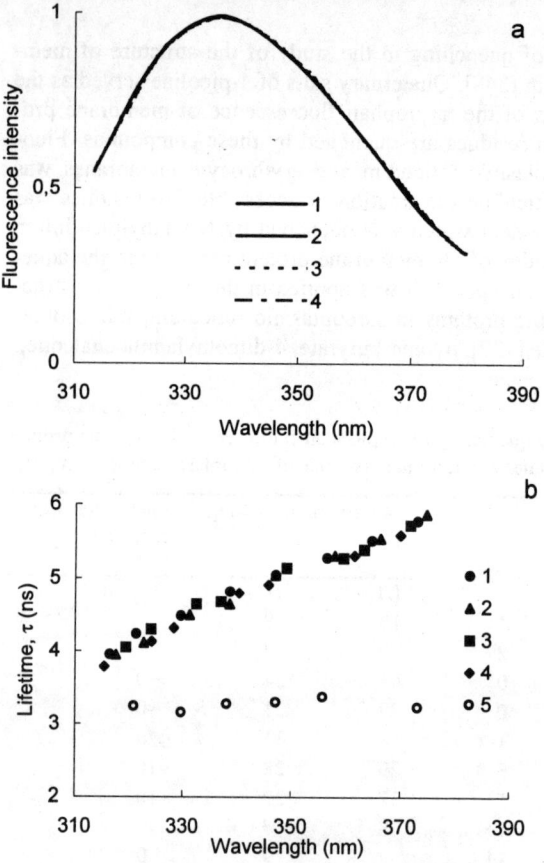

Fig. 14.1 a,b. Spectra (a) and lifetimes (b) of tryptophan fluorescence in fractions of sarcoplasmic reticulum: (1) Ca^{2+}-ATPase from longitudinal tubules, (2) Ca^{2+}-ATPase from terminal cisternae, (3) longitudinal tubules, (4) terminal cisternae, (5) lifetime of DL-tryptophan in solution. Medium: 5 mM imidazole, 100 mM sodium chloride, 2.5 mM magnesium chloride, pH 7.0 [60]

brane and the probe. For sarcoplasmic reticulum the quenching value is independent of the probe charge. The magnitude of maximum quenching characterizes the steric accessibility of tryptophan residues of membraneous proteins to the probes.

Lifetime values obtained at different emission wavelengths and fluorescence spectra for two fractions of sarcoplasmic reticulum are shown in Fig. 14.1. The fluorescence of LT and TC fractions is due to Ca^{2+}-ATPase containing 19 tryptophan residues. As known, Ca^{2+}-ATPase is the main protein component of the membranes (~80%). Tryptophan fluorescence of reticulum belongs to Ca^{2+}-ATPase (Fig. 14.1). It is known that 18 tryptophan residues of Ca^{2+}-ATPase are located in the hydrophobic fragment of the protein. This fragment is entirely buried in the membrane. Therefore, the most effective quenchers are the substances that are readily incorporated into the membranes of reticulum; for instance, pyrene, ANS, anthracene, diS-C_3-(5) (Table 14.1). The probes such as pyrene, anthracene, diS-C_3-(5) are traditionally considered as lipid probes [19]. The probes ANS and diS-C_3-(5) are used to measure transmembrane potential in membranes of reticulum [19, 349]. When the concentrations of these probes in the membrane increase, the probes reach hydrophobic regions of the protein and quench fluorescence of tryptophan. The dimensions of these probes are so large that probes could hardly penetrate inside the protein globula. Therefore, the strong quenching of the Ca^{2+}-ATPase fluorescence by probes indicates that at least the half of the tryptophan residues of this enzyme contact with the lipid phase of the membrane. This is in an agreement with the results [350] indicating that about 12 tryptophan residues of Ca^{2+}-ATPase in membranes of reticulum are accessible to N-bromosuccinimide (and ~6 are not), and with the 54 % collisional quenching by quaternary salts of 4-picoline [348]. The tryptophan residues buried inside the Ca^{2+}-ATPase molecule are sterically inaccessible to the quenchers. The value of maximum quenching is sensitive to the spatial organization of the protein. For Ca^{2+}-ATPase isolated from the TC membranes and then reconstituted into the lipid membrane, the maximum quenching by ANS was 48% (compare with data for TC in Table 14.1). Probably, in TC membranes a portion of enzyme molecules exists in the oligomerized form, where tryptophan residues are sterically less accessible for quenchers.

The data (Table 14.1) show that even at high quenching values the resonance energy transfer either has low efficiency or does not take place at all. It is most likely that the dominant process is due to "contact" deactivation by the probe which accepts vibrational quanta from the excited tryptophan residues embedded in the lipid bilayer.

14.2 Quenching of Tryptophan Fluorescence by ANS

Fig. 14.2 shows fluorescence spectra of membranes of sarcoplasmic reticulum before and after adding ANS. The quenching efficiency reaches ~40 %. The area under the spectrum of ANS emission is about 3.5 times less than that of the quenched tryptophan fluorescence. The values of the fluorescence quantum yield for the membranes and the probe are 0.3 and 0.5, respectively. The intensity of the fluorescence under the direct probe excitation at 286 nm and sensitized through

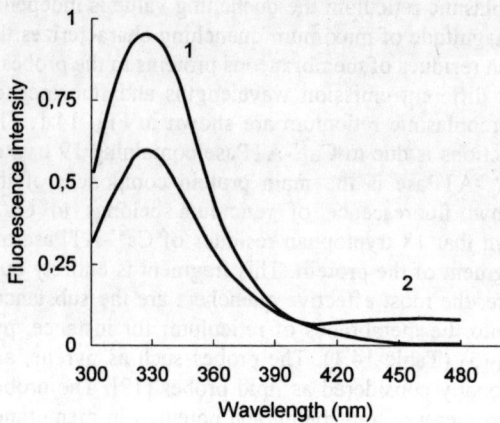

Fig. 14.2. Corrected fluorescence spectra of membranes of sarcoplasmic reticulum without ANS (1) and with ANS (2). Concentration of membranes is 0.13 mg protein / ml, of ANS is 10 µM. Excitation is at 286 nm

tryptophan excitation is 30% and 70%, respectively. The effects of screening and reabsorption were negligible. Thus, it can be concluded that energy transfer efficiency is no more than 20% of the total quenching (the spectral area the quenched fluorescence of reticulum is taken for 100%).

The prevalence of quenching over resonance energy transfer could be explained by: (a) the influence of ANS on the dynamics of quenching of tryptophan residues by neighboured groups of the protein (data on conformational changes in some proteins when they bind ANS are available [60]), and (b) quenching by ANS via collisional deactivation. The deactivation mechanism seems to be feasible, since ANS is a chromophore molecule with two deactivating groups, NH and SO_3^-.

Our results on quenching of fluorescence in membranes of sarcoplasmic reticulum by ANS substantially differ from those of Augustin and Hasselbach [351] who reported about over 90% efficiency of energy transfer. Having analyzed data represented in [351] we concluded, that the authors did not correct data on screening and reabsorption at high concentrations of ANS.

14.3 Quenching of Tryptophan Fluorescence by Pyrene

When pyrene is added to membranes of sarcoplasmic reticulum, it capable to quench tryptophan fluorescence [137]. Previously this process was explained by resonance transfer. This explanation was based on an increase in the intensity of pyrene emission and the appearance of a new band at 280 nm in excitation spectrum. However, a more thorough study disproved this explanation [60, 329].

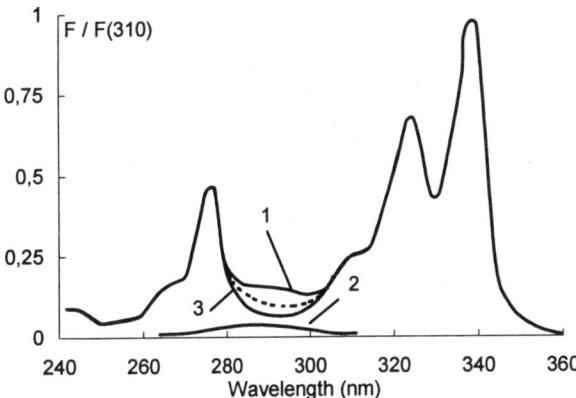

Fig. 14.3. Excitation spectra of pyrene in membranes of sarcoplasmic reticulum (1), sarcoplasmic reticulum (2), and pyrene in liposomes (3). Concentration of reticulum is 0.14 mg of protein per ml, pyrene is 4 µM, lecithin liposomes is 0.2 mg/ml. Registration is at 393 nm, slits are 4 nm, "Hitachi MPF-4", 5-mm cuvette [60]

Fig. 14.3 shows the excitation spectrum of pyrene in membranes of sarcoplasmic reticulum measured under ~ 50% quenching of protein fluorescence of reticulum. Tryptophan fluorescence of reticulum belongs to Ca^{2+}-ATPase (containing 19 tryptophan residues). A low-intensity band is observed at 280–300 nm. This band is composed of number components. These components are due to: (a) protein fluorescence not quenched by pyrene at 393 nm; (b) a direct excitation of pyrene; (c) emission of pyrene induced by small reabsorption of protein fluorescence; and (d) very small resonance transfer (the area below the dotted line in Fig. 14.3). If at least nine tryptophan residues of the enzyme could participate in resonance transfer, the 280-nm band should be many times stronger than the mentioned area (it was calculated using extinction coefficients of tryptophan and pyrene).

Thus, it is obvious that no efficient resonance transfer takes place. This means that the quenching of protein fluorescence in reticulum by pyrene should be explained by other processes. Actually, quenching could be induced by deactivation or exciplex formation.

14.4 Tryptophan–NADH Pair in Mitochondria

A small change in the fluorescence intensity of proteins in mitochondrial membranes was observed during the transition from one functional state to another caused by the addition of succinate or ADP [18, 352–354]. Many researchers attribute these changes to conformational transitions in membranous proteins [358].

However, the changes in the fluorescence intensity of proteins in mitochondrial suspensions are opposite to the changes in the fluorescence intensity of mitochondrial NADH [354]. Therefore, it was suggested that these changes are a result of the energy transfer from tryptophan residues to NADH.

We repeated the experiments from [354] with some modifications. Fluorescence intensities of mitochondrial proteins and mitochondrial NADH in hypotonic and isotonic media are shown in Fig. 14.4 [355].

Unlike the emission of NADH in water [205], the emission of mitochondrial NADH has maximum at 440 nm, which corresponds to the protein-bound NADH. The lifetime of mitochondrial NADH is 1.8 ns. This value coincides with that of NADH bound to alcohol dehydrogenase. The lifetime of NADH in aqueous solution is about 0.5 ns. When the fluorescence intensity of mitochondrial NADH decreases, the lifetime does not change. Therefore, the decrease in fluorescence intensity is due to the conversion of NADH to NAD^+, but not due to desorption of NADH or a quenching process. It may be also assumed that all mitochondrial NADH molecules are bound to proteins.

We also measured excitation spectrum of mitochondrial NADH (Fig. 14.5). An intense tryptophan band at ~290 nm and an NADH band at 340 nm were observed

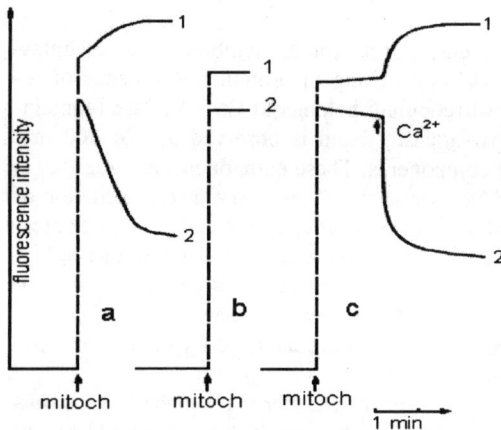

Fig. 14.4. Changes in the fluorescence intensity of mitochondrial proteins and NADH in different media: a) in a hypotonic solution (5 mM Tris-HCl, 4 mM potassium phosphate, pH 7.5; b) in the hypotonic solution containing 0.5 mM EGTA; c) in an isotonic solution (5 mM Tris-HCl, 4 mM calium phosphate, 50 mM potassium chloride, 150 mM sucrose, pH 7.5). The concentration of rat liver mitochondria was 0.3 mg protein per ml. Calcium chloride was added in c) at concentration of 20 µM. Curves: 1) excitation at 286 nm, emission at 330 nm, 2) excitation at 340 nm, emission at 440 nm

The spectrum was similar to that of NADH bound to alcohol dehydrogenase.

Thus, the increase in the fluorescence intensity of mitochondrial proteins observed in the hypotonic medium or after calcium addition results from disappearance the tryptophan-NADH energy transfer when NADH converts to NAD^+.

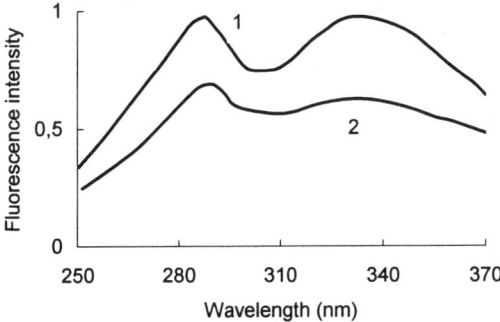

Fig. 14.5. Excitation spectra of mitochondrial NADH in a suspension of native mitochondria before (1) and after (2) adding calcium. Emission is monitored at 440 nm, slits are 4 nm, "Perkin-Elmer MPF 44B", without correction for spectral sensitivity and scattering. Concentration of mitochondria is 0.05 mg protein/ml; concentration of added calcium is 80 µM

14.5 Photosynthetic Reaction Centers

The antenna effect, i.e. a migration of energy along identical or similar chromophores to special chromophore traps, occurs in natural photosynthetic structures such as chloroplasts in plants, chromatophores and phycobilisomes in bacteria [38, 39, 356–363]. The antenna effect allows to use the energy of sunlight with maximum efficiency to maintain vital biochemical reactions. The antenna effect can be demonstrated with the example of bacterial pigment-protein complexes in chromatophores and phycobilisomes.

One of the most frequently occurring pigments in chromatophores is bacteriochlorophyll, a tetra-pyrrol porphin compound with a magnesium ion at the center. Most bacteriochlorophyll molecules are giant antennae which collect light and transfer absorbed energy in the form of a singlet electronic excitation to chromophore traps, i.e. reaction centers where charge is separated [38, 39, 363]. According to X-ray analysis data, the light-harvesting unit isolated from *Prosthecochloris Aestuarii* bacteria is bacteriochlorophyll-protein noncovalent complexes where protein is composed of three subunits, each containing 7 pigment molecules. The average distance between the centers of neighbour pigments is about 12 Å, and between the pigments of different subunits is about 24 Å. The maximum of the long-wavelength absorption band is ranging between about 700 and 900 nm (the extinction coefficient is about 100 000 $M^{-1}cm^{-1}$), depending on the type of bacteriochlorophyll.

It has been found that energy migration between bacteriochlorophyll molecules occurs very rapidly, no longer than 1 ps for one migration step. Due to such very high transfer rate the processes of deactivation, fluorescence, etc. are unable to compete for the population of the excited molecules. The diffusion of molecules does not play a significant role in energy migration.

Migration and flow of excitation in chromatophores and phycobilisomes is generally interpreted either in terms of Forster theory or in terms of exciton interactions, or by both theories. The Forster model is usually applied in model studies on energy migration between porphyrin rings incorporated into globins.

Subpicosecond energy transfer within various light-harvesting antenna mutants of purple bacteria (*Rhodobacter sphaeroides*) at 77 K was studied in [364]. When the spectral overlap between the 800-nm band and the variable bands increases, the rate of energy transfer measured by the lifetime of the 800-nm band increases. The authors state that the rate of energy transfer qualitatively agrees with the Forster spectral overlap calculations. Figures in [364] show the dependencies of energy-transfer time from ~1 to 2 ps. Such a rate is more likely for the hot transfer, rather than for cold transfer described by the Forster model.

The following arguments are against the traditional approach using Forster theory: a) Interpigment distances in chromatophores and phycobilisomes are relatively small. For instance, the *X*-ray analysis of the crystals of reaction centers of *R.viridis* bacteria demonstrated that distances between the centers of the pigments (bacteriochlorophyll and bacteriopheophytin) are ranging between 11 and 13 Å and the distances between chromophore edges are within 3 and 5 Å. The size of the pigment molecules is about 15Å. The size of these chromophores were even larger than the distances. Similar values can be found for other bacteria. (b) Measurements of absorption, circular dichroism and optical rotatory dispersion spectra show a splitting in the spectra. The splitting points to a relatively strong exciton interaction of pigments in chromatophores and phycobilisomes. (c) The migration rate is very high, so that after absorption of a photon the donor molecule has no time to transform into an equilibrium state. The rate of migration is about 10^{12} s^{-1} that is at least 100 times higher than the theoretical Forster rates. Since energy transfer between chromophores of photosynthetic reaction centers occurs on the picosecond and sub-picosecond time scales [38, 39, 364, 365], this transfer should be hot, and in this case the use of Forster theory is incorrect.

14.6 Conclusion

Efficient energy transfer in biomembranes occurs over short distances comparable with chromophore size. The efficiency of electronic excitation energy transfer is usually less than the efficiency of quenching due to competing processes. The quenching of donor fluorescence is often the result of its deactivation via dynamic or static mechanisms. Forster theory is inadequate for the quantitative description of energy transfers in the above systems.

The changes in the fluorescence of mitochondrial proteins during transitions between different functional states result from the cessation of energy transfer from tryptophan residues to protein-bound NADH, which is converted to NAD$^+$.

15 Fluorescence Probes

15.1 Widely Used Probes

In subcellular biochemistry, biomembrane structure and dynamics are widely investigated with fluorescent probes [366]. Many probes sensitive to pH, polarity, calcium, electric potential, etc. are described in the handbook of fluorescent probes [367]. The spectroscopic properties of some widely used fluorescent probes are shown in Fig. 15.1.

One of the most widely-used probes is pyrene, aromatic hydrocarbon. Pyrene has very long natural lifetime, about 300 ns, that is the reason of its high sensitivity to a lot of chemical compounds, rigidity of the microenvironment, temperature, etc. Pyrene and its derivatives are used as probes and labels in the studies of membraneous viscosity and phase transitions, enzyme flexibility and protein complex formation, nucleic acids structural investigations, etc. [263, 330, 368–382] (Chapters 14 and 16).

For instance, it was shown [379, 380] that the 5'-pyrene-labeled RNA oligomers is unusually attractive, stable and very sensitive probes for studies of oligomer-oligomer and oligomer-ribozyme interactions; they have minimal perturbations on the thermodynamics of secondary and tertiary structure formation in RNA [379]. Hydrolysis of a pyrene-labeled ATP to Pyr-ADP by myosin subfragment-1 was accompanied by an initial quenching of the pyrene fluorescence with a subsequent recovery of the fluorescence [381].

Steady-state and phase-modulation fluorescence methods for calcium-sensitive probes (QUIN-2, etc.) were used in [383, 384]. The kinetics of fluorescence probes (probes with a large Stokes shift, probes sensitive to intramolecular quenching process, probes for biology – 1,8-ANS, 2,6-TNS, DNS-C1, PRODAN, etc.) were studied using synchrotron radiation in [385]. The fluorescent probe laurdan was used to investigate structural organization of the vesicular membranes of the stomatitis virus in [386].

The fluorescence of 9-aminoacridine is quenched in vesicular suspensions containing negatively charged lipid headgroups (e.g. phosphatidylserine) by a transmembrane pH-gradient [387]. This fluorescence quenching is accompanied by the formation of 9-aminoacridine dimers, which in the excited state are transformed into an excimer state. This result has been obtained based on specific dimer excitation spectrum and hypochromic absorption spectrum that are red-shifted

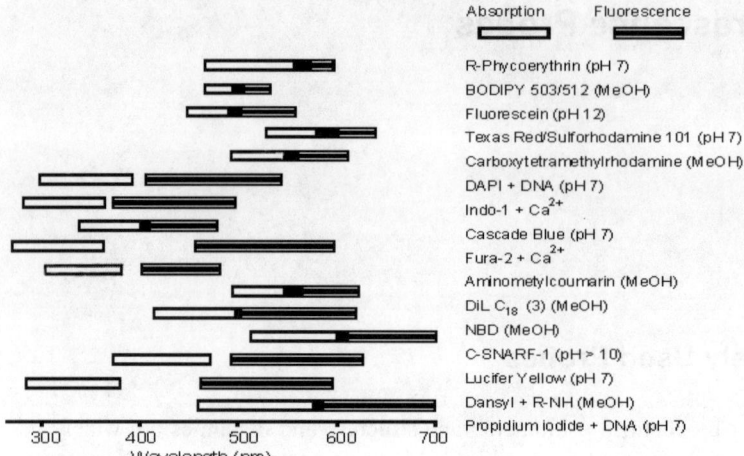

Fig. 15.1. Spectral ranges of absorption and fluorescence for 16 fluorophores of most importance. The range covers only those wavelengths where the values of absorbance or fluorescence intensity exceed 25% of the peak value [367]

by no more than 275 cm^{-1}, as well as on the detection of the characteristic broad excimer emission band, centered at 560 nm [387].

The pulsed laser photoacoustic technique gives excellent results in the measurements of fluorescence quantum yields of fluorophores (prodan, dansyl, acrylodan) in solution, but anomalous results for fluorophores bound to proteins [388]. The time-resolved photoacoustic spectroscopy method was considered in [389].

In [390, 391] Zn-mesoporphyrin IX served as a probe on conformational fluctuations in apomyoglobin at temperatures as low as 2 K. The stimulated photon echoes method was used. The sample was excited by three laser pulses (duration of ~5 ps, bandwidth of ~8 cm^{-1}) that overlap in space but do not overlap in time. Their frequency was resonant with an optical transition of the probe. The first two pulses generated a frequency grating within the spectrum of the probe. A frequency grating is a modulation of the probe population in the ground state and in the excited state as a function of their resonance frequency. Just as an optical pulse is scattered from a spatial grating, the third pulse is scattered from the frequency grating as the photon echo. The intensity of the echo signal is proportional to the modulation depth of the frequency grating. A similar approach was applied to Zn-substituted cytochrome c [390, 392]. The observed conformational dynamics was interpreted as a spontaneous mobility in the proteins However, a photoinduced mobility also should be taken into account in such experiments.

One of the first applications of polarized fluorescence depletion was described in [393] for individually selected mammalian cells. This technique combines the sensitivity of fluorescence detection and the long lifetime of a triplet probe (eosin), to measure rotational diffusion of a membrane protein with ~0.1 ms time. Measuring the differences in the rates of donor photobleaching in the cells labeled with donor only (fluorescein isothiocyanate-conjugated proteins) and with donor

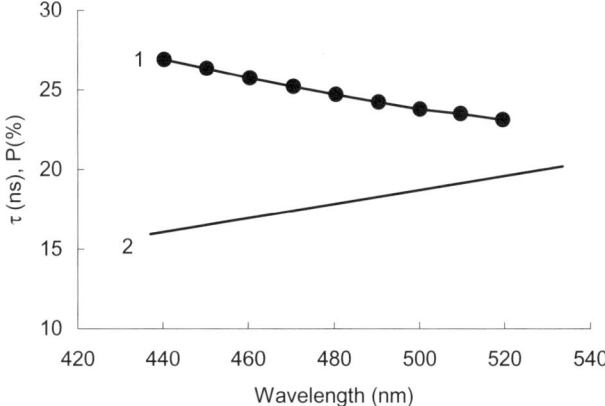

Fig. 15.2. Fluorescence polarization degree (1) and average phase-modulation lifetime (2) of ANS bonded with BSA (Sigma, 5-th fraction). Concentrations are: BSA is 15 µM, ANS is 12 µM. Excitation is 370 nm. Life-time measurements were performed with SLM-4800 at 18 MHz [525]

and acceptor (tetramethylrhodamine-conjugated proteins) allows to calculate the efficiency of energy transfer [394]. Fluorescence recovery after photobleaching with evanescent interference patterns was to measure the translational diffusion coefficients of the weakly bound fragment 1 of bovine prothrombin and of fluorescent lipid NBD-PC for substrate-supported planar membranes composed of lipids [395]. The method of total internal reflection illumination in fluorescence microscopy with photobleaching recovery and evanescent field polarization was reviewed in [396].

Fluorescence polarization anisotropy of intercalating dyes is used in investigation the structure and orientations in nucleic acids. For instance, the authors [519] developed and experimentally verified an expression of the anisotropy function for short DNA duplexes with intercalated ethidium bromide. Modeling the eight duplexes as cylinders, they calculated a duplex diameter of ~19 Å.

A simple method for sizing pores in vesicles based on the differential release of co-encapsulated fluorescein-labeled dextran markers of two different sizes was reported in [397]. The method was tested using the bee venom peptide mellitin, which was found to form pores of 25–30 Å diameter in palmitoyloleoylphosphatidylchohne vesicles.

In [398] Monte Carlo simulations of fluorescence recovery after photobleaching experiments on two-component lipid bilayer systems in the solid-fluid phase coexistence region were carried out to study the geometry and size of fluid domains in these bilayers.

Fluorescence properties of one of the most widely used probes, 1-anilmonaphthalene-8-sulfonate (ANS) are represented in Table 15.1. According to the polarization data (see also Table 15.1), ANS in biomembranes is preferentially

Table 15.1. Fluorescence of ANS in different systems

System	Q_f	τ, [ns]	P	F_{max}, [nm]
Water	0.004	0.55	>0.01	510
Ethanol	0.37	7.7	~0.0	468
Lecithin layers	~0.3	~7.5	~0.1	475
Mitochondria	0.1	8.7	>0.3	480
Sarcoplasmic reticulum	>0.5	10.0	>0.3	475
Erythrocyte shadows	-	~10.0	0.3	470
Bovine serum albumin	~0.6	15.9	>0.3	468
Apohemoglobin	0.9	10.4	0.3	457

Q_f is the fluorescence quantum yield, τ is the lifetime, P is the degree of polarization, F_{max} is the wavelength of the emission maximum. The data are borrowed from [19, 60, 399, 400]; the values shown here were obtained at low ANS concentration, 20°C, under aerobic conditions.

fluorescens from proteins, but not from lipids. ANS is fixed more rigidly on proteins than in lipids. ANS bounded with proteins has poor sufficient rotational mobility. The value of the polarization degree varies a little depending upon the type of protein.

In most cases, the fluorescence of probes in biostructures is heterogeneous: the emission maximum depends on the excitation wavelength; the fluorescence decay is non-exponential; lifetime and polarization are changed with the emission wavelength. The spectral emission heterogeneity may be caused by two reasons: (a) heterogeneity of the absorption centers; (b) heterogeneity of the emission cen-

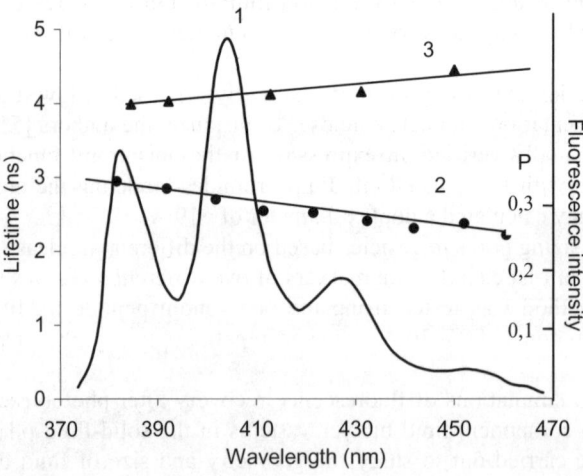

Fig. 15.3. Fluorescence spectrum (1), polarization (2) and the lifetime (3) of anthracene bound by bovine serum albumin. Excitation wavelength is 360 nm. The concentration of albumin is 14 μM, of anthracene is 4 μM

ters. If a probe molecule in a biostructure is surrounded by a number of different groups, a number of different excited-state interactions can arise after photoexcitation. Therefore, a number of emission centers could be detected for the population of probe molecules. In many cases, the spectral emission heterogeneity can arise without any absorption heterogeneity. An emission heterogeneity in polarization and lifetime is shown in Fig. 15.2 for ANS bonded with bovine serum albumin. The "mirror" behaviour in $P(\lambda)$ as compared with $\tau(\lambda)$ is a typical situation for probes in biopolymers. In many cases, the $P(\lambda)$ courve could be calculated using the $\tau(\lambda)$ courve according to the Levshin-Perrin equation (15.1). The increase in τ with the emission wavelength is due to increase in the amplitude of the long-lived lifetime component during an excited-state relaxation.

Fig. 15.3 shows the fluorescence spectrum of anthracene bound to bovine serum albumin. Unlike ANS, the photoexcited anthracene is not a static dipole. However, the spectral behavior of the lifetime and polarization is basically the same as for ANS and for tryptophan in proteins. The spectral heterogeneity of anthracene emission is due to number types of the excited-state interactions with different amino-acid residues surrounding one anthracene molecule in BSA.

15.2 Estimation of Sizes of Chaperones and their Complexes Using ANS

SecB is a molecular chaperone, a homotetrameric protein that forms a specific complex with precursor proteins. *In vitro,* SecB is capable of binding to a variety of unfolded proteins, but not to native proteins [401]. The molecular weight of the SecB tetramer is 66.4 kDa [402]. SecB may be isolated from *E.coli* by conventional method. *In vitro* the complex formation between SecB and nonnative polypeptide was detected using the acrylodan fluorescence from the labeled polypeptide [403]. However, the acrylodan labeling is a complicated difficult method.

In work [402], ANS was used for SecB probing and steady-state fluorescence detection of the SecB-polypeptide complex formation. ANS binds to the hydrophobic region of SecB. When this probe is bound, the intensity of fluorescence increases and the emission maximum is shifted from ~520 nm to shorter wavelengths. Upon adding nonnative polypeptides to SecB the emission is increased and shifted to 472 nm. In a collaborate work with P.Fekkes and A.Driessen (University of Groningen), we have developed the Randall's approach [401,402] in conjunction with polarization and time-resolved spectrofluorimetry to analyze the complex formation between polypeptides and SecB [525]. As a polypeptide model, we have used a good substrate for SecB - the reduced form of bovine pancreatic trypsin inhibitor (RBPTI).

BPTI is a protein with a molecular mass of 6 kDa. It has not tryptophan residues but contains tyrosines and three disulfide bonds that stabilize the tertiary structure. By blocking the reduced dithiols with iodoacetamide, the protein adapts a random conformation without any pronounced native-like secondary structures.

SecB contains 4 tryptophan residues, 1 tryptophan per 1 subunit of the tetramer

SecB. Tryptophan emission spectrum of SecB has a maximum at ~340 nm. The addition 10 µM of RBPTI to 3 µM of SecB changes the signal at 340 nm less than 10%. This ~10% consist of four small components: (a) the SecB tryptophan fluorescence changes; (b) the tyrosine emission of RBPTI; (c) penetration of the exciting light into a registration channel; (d) Raman scattering. The quantitative separation a) from b), c) and d) is hardly possible. No any control experiments on the nature of the signal are described in [401, 403]. In [401] the signal was detected with too large emission bandwidth, 20 nm.

Binding of SecB with ANS. The SecB tryptophan fluorescence changes hardly could be a test on the SecB-RBPTI complex formation. Therefore, according to [402], we have used the environmental sensitive fluorophore ANS. When excited at 370 nm, ANS emits between ~450 and ~550 nm. Upon SecB addition to the ANS solution, the increasing of intensity and the blue shift of the ANS spectrum from 520 to 490 nm were observed (Fig. 15.4). These changes take place due to the ANS binding with the protein. The ANS emission in the presence of SecB was not changed upon DTT addition.

When ANS binds with the proteins (Table 15.2), ANS polarization degree sharply increases, because a large probe, the size > 10 Å, may hardly rotate itself in the bound state. In this case, polarization degree depends mainly on the rotation of the probe together with the protein globule as a whole. For ANS in water, a small polarization degree is due to a rapid freedom rotation.

Polarization degree and modulation lifetime of ANS bonded with SecB are dec-

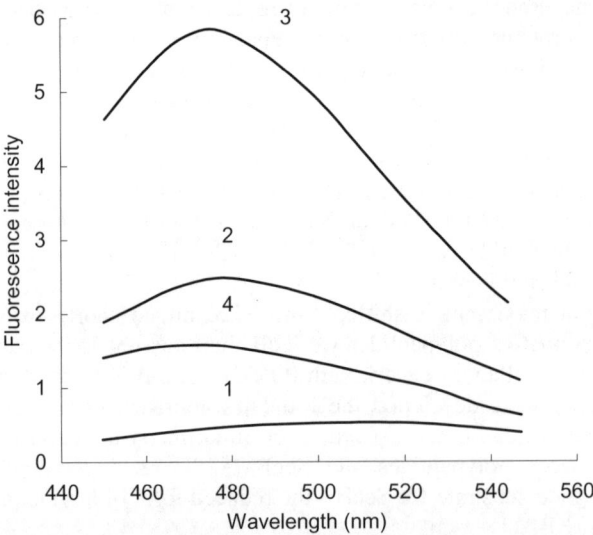

Fig. 15.4. Emission spectra of ANS in buffer (1), with SecB (2), with SecB + RBPTI (3), and with RBPTI (4), in the presence of DTT. Concentrations: ANS is 3 µM, SecB is 18 µM, RBPTI is 107 µM, DTT is 50 mM. Excitation is 370 nm, slits are 4 nm; mirror microcuvettes were used

15.2 Estimation of Sizes of Chaperones and their Complexes Using ANS

Table 15.2. Estimation of the volume sizes and diameters of SecB, RBPTI and the SecB-RBPTI complex by lifetime and polarization degree of ANS

Parameter	ANS	ANS+SecB	ANS+RBPTI	ANS+SecB+RBPTI
P	<0.06	0.197	0.127	0.237
τ [ns]	3.8	6.2	8.0	14.0
V [Å^3]	<2260	19550	12376	64729
d [Å]	<16.0	33.4	28.7	49.8

The proteins were dissolved in 20 mM Tris-HCl buffer (pH 7.6) at 25°C. ANS concentration was 12 µM. Excitation was 370 nm, emission was 500 nm. For increasing of a signal in average lifetime measurements, the mirror cuvettes were used. Modulation frequency was 18 MHz.

reased from the «blue» region of an emission spectrum to the «red» one (not shown). No a «mirror» ratio between polarization degree $P(\lambda)$ and lifetime $\tau(\lambda)$ along the emission spectrum was observed for this protein. Such "non-mirror" picture indicates the existence of at least two ANS emission centers; the discussion about correlation of $P(\lambda)$ and $\tau(\lambda)$ see in [60, 69, 167, 183]. The decay was sufficiently non-exponential; phase life-times were much smaller than the modulation ones: ~1 ns of a small amplitude and 6 ns of a large amplitude, respectively. It can be concluded that at least two ANS emission centers exist in SecB: (a) the short-lived center (~1 ns) where ANS is strongly quenched by some polar amino-acid residues, (b) the long-lived center (6 ns) where ANS emits from the hydrophobic pocket. These data confirms the suggestion in [402] about existence in SecB two sites with different binding requirements.

The tryptophan emission is overlapped with ANS absorption; therefore, the resonance energy transfer could be expected. However, a quenching of the tryptophan SecB fluorescence was not observed (only screening and reabsorption took place in the experiment). Thus, ANS is bound in the hydrophobic region far away from tryptophan residues of the tetrameric globule.

RBPTI and ANS. The ANS emission spectrum in aqueous solution is changed upon RBPTI addition (Fig. 15.4). Polarization degree and life-time are changed drastically on the binding (Table 15.2). Lifetime of ANS with RBPTI was a bit more than that with SecB. However, the fluorescence intensity was sufficiently less than with SecB. No any influence of ANS on the RBPTI tyrosine fluorescence was observed.

SecB-RBPTI complex. ANS-labeled SecB binds to RBPTI with a considerable increase in probe fluorescence and with a shift in the emission maximum from 490 to 480 nm (Fig 15.4). The blue shift and increase in fluorescence quantum yield upon binding of RBPTI is indicative of a transfer of the ANS either to a more unpolar environment or (and) to SecB conformational changes. A relatively high quantum yield of ANS in the SecB-RBPTI complex comparing with a probe emission from SecB or RBPTI separately, allows to apply this approach for detection of the complex. The best result is achieved with low content of ANS in comparing to the proteins.

The complex formation leads to an increase in τ (Table 15.2). Polarization degree and the modulation life-time at 500 nm for ANS with SecB, RBPTI and SecB-BPTI complex are represented in the same Table. The life-time decays were non-exponential. The modulation life-time component decreases, from "blue" to «red» region.

Estimation of the protein sizes. The known Levshin-Perrin equation gives the ratio between lifetime (τ), polarization (P), temperature (T), viscosity (η), and the volume of a particle (V). This equation may be modified to the view:

$$V = (1/P_0 - 1/3) R T \tau / (1/P - 1/P_0) \eta \qquad (15.1)$$

Here P_0 is the limit polarization, R is the gas constant. Using the obtained equation, V and diameters (d) of proteins and their complexes can be calculated. If $T=300$K, $\eta=10^{-3}$ kg m^{-1}s^{-1}, $R=8.3$ kg m^{2}s^{-2} K^{-1} mol^{-1} and $P_0=0.42$, we obtain:

$$V = 8498\, \tau / (1/P - 2.381) \qquad (15.2)$$

Here τ expressed in nanoseconds, V is in Å^3. The calculations for SecB, RBPTI and the SecB-RBPTI complex by P and τ of ANS at 500 nm (at 370-nm excitation) are represented in Table 15.2. The P_0 value 0.42 was chosen from the data on polarization of a dilute solution of ANS in propylene glycol at –55°C [399]. As seen from the Table, the volume size of the complex is much more than the sizes of SecB and RBPTI separately. The volume ratio of RBPTI and SecB in the complex is equal ~3.7, i.e., about 4 molecules of RBPTI per 1 SecB globule (tetramer). This result agrees with known [403] molecular proportion in the SecB-RBPTI complex.

15.3 Fluorescent Studies of Na$^+$ K$^+$-ATPase

The Na$^+$ K$^+$ ATPases use energy from the hydrolysis of ATP to transport Na$^+$ and K$^+$ ions. The mechanism of the Na$^+$K$^+$ATPases is far from being fully understood. One of the main problem is an experimental detection of various conformational states of the enzyme upon ATP and cation binding [404–408].

Since Na$^+$ K$^+$ ATPases has many tryptophan residues, their emission was applied for detection of conformational states of the enzyme [405]. The protein fluorescence was changed upon ATP and ion binding. It was shown that ATP accelerates the rate of conversion E2 (K$^+$) to E1 (K$^+$) [404, 405]. Unfortunately, the amplitudes of the fluorescence response in the experiments were too small, a few percents.

More pronounced fluorescence responses were obtained with probes and labels. Eosin, 5-iodoacetamidofluorescein, formycin nucleotides [404], fluorescein isothiocyanate [406] and potential-sensitive styryl-based dyes [407, 408] were used. The most sensitive to enzyme transitions were zwitterionic styryl dyes RH160, RH421, and also di-4-ANEPPS [408]. In membranes with Na$^+$K$^+$ATPases the

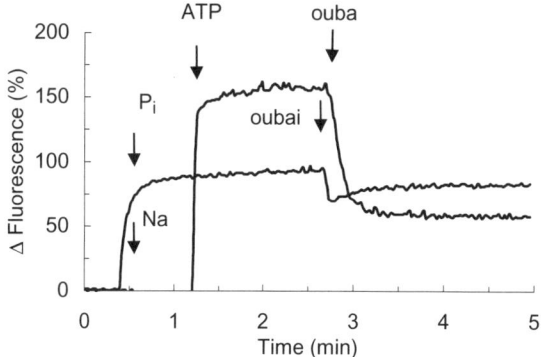

Fig. 15.5. The RH421 fluorescence changes in membranes with Na$^+$K$^+$ATPases (in the presence of 4 mM MgCl$_2$) [408]. Additions are: 16 mM of NaCl, 30 μM of ATP or 4 mM of inorganic phosphate, 20 μM of ouabain. Excitation at 580 nm, emission at >630 nm. Concentration of RH421 is 0.4 μM [407]

dyes changes the absorption and emission upon ATP and cation addition. Using RH421, unphosphorilated and phosphorilated states of Na$^+$K$^+$ATPases from rabbit kidney were detected in [407]. Typical recording of RH421 fluorescence changes caused by enzyme phosphorylation and following ouabain binding is shown in Fig. 15.5.

According to the initial hypothesis, the RH dye is inserted into the lipid domains of the membrane and detects changes of electric field, but not of the E1-E2 conformational transitions of the enzyme [407]. According to another point of view [483], styryl dyes localized in close proximity of Na$^+$K$^+$ATPases and interact with it; this interaction, detected by fluorescence, mainly depends on the enzyme charges and conformation, but not on the membranous electric field.

Measurements of energy transfer between site-directed fluorescent probes were applied to calculation the distance between the ATP site and the ouabain-binding site [404]. Both fluorescein isothiocyanate and TNP-ATP were used as probes for the ATP site, and anthroyl ouabain and several derivatives of digitoxigenin were used as probes for ouabain site [404]

15.4 Anthracene with Dimethylaminochalcone in Membranes

When different chromophore compounds are incorporated into membranes, one compound may quench another. For instance, the fluorescence of aromatic hydrocarbons can be quenched by dyes, etc. [19, 20, 60, 409]. Usually this phenomenon is also considered in terms of inductive resonance. However, the deactivation process and exciplex or excimer formation can be dominant.

Several anthracene derivatives were taken as donors of electronic excitation for the 4-dimethylaminochalcone dye in the lipid membrane [409, 410]. The dye efficiently quenched anthracene fluorescence. Fluorescence and absorption spectra of anthracene and dimethylaminochalcone strongly overlap, so that R_0 ~40Å [409, 410]. The authors substituted the efficiency of quenching in a modified Forster equation and obtained values of intermolecular distances that characterized the depth of "embedding" of the probes in the membrane with the accuracy of ~1 Å.

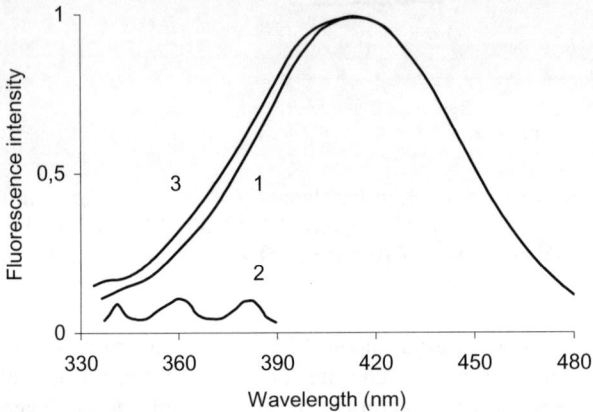

Fig. 15.6. Excitation spectra of 4-dimelhylaminochalcone (1), anthracene (2) and both substances (3) in liposomes. Spectrum (3) was recorded when the efficiency of quenching of anthracene by dimethylaminochalcone was about 50% (point A on Fig. 15.7). The emission is monitored at 520 nm, the slits are 2 nm, "Hitachi MPF-4". The concentrations: anthracene is 5.6×10^{-6} M, dimethylaminochalcone is 1.1×10^{-5} M, liposomes is 0.6 mg/ml

However, the size of the probes should be taken into an account. These are large anisotropic molecules. The size of anthracene and dimethylaminochalcone is about 7 and 12 Å, respectively. Therefore, an accuracy in estimation of the distances hardly can be 1 Å. The probes may be unevenly distributed throughout the membrane and contact each other. Moreover, the diffusion-dependent collision is also possible. Let us try to clarify the actual mechanisms of quenching.

Fig 15.6 shows excitation spectra of 4-dimethylaminochalcone, anthracene, and of the combination of both probes in lecithin liposomes where the quenching efficiency was about 50% (Fig. 15.7). No anthracene band is seen in the excitation spectrum of dimethylaminochalcone. A small component in spectrum 3 is due to the intrinsic anthracene fluorescence. The fluorescence of 4-dimethylaminochalcone is not sensitized. Thus, the experimental results show no significant transfer of electronic excitation from the anthracene chromophore to dimethylaminochalcone.

As known, anthracene and its derivatives form fluorescent exciplexes with diethylaniline or dimethylaniline [57]. 4-dimethylaminochalcone is an analog of dimethylaniline:

15.5 Diffusion of Probes

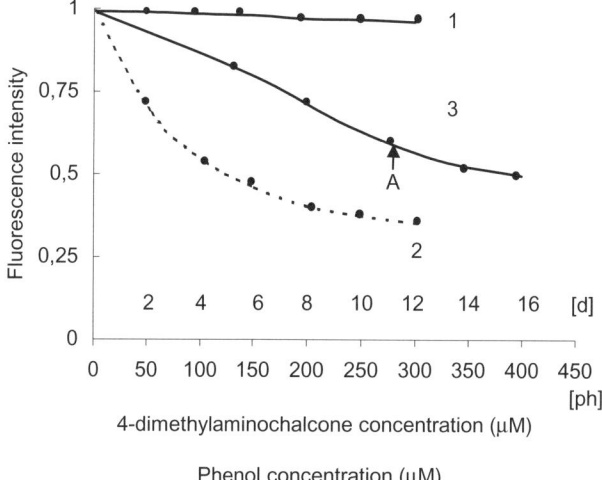

The formation of the anthracene-dimethylaminochalcone exciplex depends on the same groups which are involved in the anthracene-dimethylaniline exciplex. These groups are nitrogen with the unshaved electron pair and two methyl groups. In polar media the quantum yield of the exciplex emission is very low so that the exciplex luminescence cannot be detected because of the intense fluorescence of dimethylaminochalcone and anthracene. Dimethylaminochalcone decreased fluorescence lifetime of anthracene. The observed dynamic diffusion-dependent quenching seems to be of exciplex character.

Fig. 15.7. (1) Dependence of anthracene fluorescence intensity in liposomes on the phenol concentration ([ph]). Concentrations: anthracene is 8.7×10^{-6} M, liposomes is 0.2 mg/ml. Excitation is 378 nm, emission is 405 nm. (2) Dependence of 4-dimethylaminochalcone fluorescence intensity in liposomes on the phenol concentration. Concentrations: 4-dimethylaminochalcone is 5.6×10^{-6} M, liposomes is 0.2 mg/ml. Excitation is 420 nm, emission is 520 nm. (3) Dependence of anthracene fluorescence intensity on the concentration of 4-dimethylaminochalcone ([d]). Concentrations: anthracene is 5.6×10^{-6} M, liposomes is 0.6 mg/ml. Excitation is 378 nm, emission is 405 nm

Anthracene fluorescence in liposomes insignificantly changes with adding phenol (Fig. 15.7). Phenol is a polar molecule solubilized in water. It not penetrates into hydrocarbon region of the phospholipid membrane where anthracene is localized. However, phenol can penetrate into the polar region of a phospholipid bilayer. It quenches the dimethylaminochalcone fluorescence (Fig. 15.7), because the dye is localized in the polar region. Phenol has not an absorption band in the visible region, i.e., resonance energy transfer from dimethylaminochalcone is impossible. Quenching of the dimethylaminochalcone fluorescence by phenol is due to the dynamic deactivation.

15.5 Diffusion of Probes

The fluorescence lifetime (τ) for anthracene in liposomes is 4.6 ns. For anthracene derivatives τ is known to be approximately 9–13 ns [19, 411]. The main path length (X) for a molecule with effective radius b in a medium with a viscosity η at a temperature T during the time τ could be calculated using the equation [268]:

$$\Delta X^2 = \frac{RT\tau}{3 \times 3.14 \eta b} \qquad (15.3)$$

Substituting the values of all parameters in this equation gives ΔX=6 and 10 Å for dimethylaminochalcone and anthracene, respectively, in a lipid bilayer; hence the diffusion of probes is significant. Using fluorescence quenching by phospholipid with a spin-label attached to its polar headgroup, was shown that in the excited state anthroyloxy-labeled fatty acids move toward the surface (polar region) of the bilayer [346].

Three types of probe diffusion in membranes and biopolymers are known (mobility can be lateral and rotational): (a) Normal thermal spontaneous mobility. Anisotropy of macromolecules and membranes leads to a "sifting" of probes in certain direction. For instance, the lateral diffusion coefficient for pyrene in dipalmitoyllecithin bilayers at 50°C equals 1.4×10^{-7} cm^2 s^{-1}, that is about 10 times higher then the lateral diffusion coefficient for dipalmitoyllecithin (1.8×10^{-8} cm^2 s^{-1}) [174]. Using (15.3) and Table 15.1, it is easy to calculate the mean path length X for ANS molecule in a lipid bilayer. With $\tau \sim 8$ ns, $b \sim 5$ Å, $T \sim 300$ K, $\eta \sim 0.1$ poise (diffusion viscosity along lipid chains), the value of X will be about 10 Å. Smaller molecules may cover longer distances. (b) Charge-induced mobility: the dipole moment of a chromophore usually increases as a result of photoexcitation. This leads to an increase in the Coulombic interactions with the microenvironment. For example, the dipole moment of ANS is about 5 D in the ground state and 40 D in the excited electronic state [18]. This means that photoexcited ANS will start moving in the membrane to the regions of higher polarity. (c) Induced thermal mobility. This is due to the short-time heating of the microenvironment of the chromophore in the course of vibronical relaxation after photoexcitation (in Chapter 8).

15.6 Fluorescence Pharmacology *in vitro*

Fluorescent analysis is widely applied for testing pharmacological compounds in biostructures [400, 412–416]. For instance, cholinergic fluorescence probes were used in study of the receptor proteolipids [412]. ANS was applied to biomembranes as probe on conformational transitions induced by drugs [400]. Barbiturates decreased the amount of binding sites of ANS with lipid or erythrocyte membranes [417]. Using dansyl-derivatives, auramine, and ANS, the fluorimetric characterization of two specific drug binding sites on human serum albumin was made in [413]. Spectroscopic and photochemical studies of drugs used in photodynamic therapy is reviewed in [416]. Flow cytofluorimetry is used for drug screen [415]. Fluorescent probes may be used in the drug screening and prediction of activities of pharmacological compounds [400, 414, 415, 417].

Entering a drug in organism leads to the drug redistribution in membranes and other structures of cells, and then to a contact interaction of them with some "receptor" (enzyme, nucleic acid, etc.) located in given cellular structure. Pharmacological compound induces beginning of physiological reactions. Therefore, the pharmacological compounds should consists of: (a) the hydrophilic polar fragment, allowing the drug to be dissolved in blood and then in the aqueous phase of cells; the hydrophilic specific fragments and polar groups are responsible for the contact and influence on the receptor, (b) the hydrophobic unpolar fragment, allowing the drug to be successfully draw from the aqueous phase to cellular structures and to be accumulate in them.

The fluorescent characteristics of pharmacological compounds are straightly connected with the amount and location of their polar and unpolar fragments and groups. These fragments and groups cause their ability to quench or stimulate fluorescence [233].

Therefore, a correlation between the pharmacological activity and fluorescence parameters should be observed. High correlation means a possibility of successful fluorescence screening of medicines and prediction of their cellular targets. Many years ago Szent-Gyorgyi observed that fluorescence quenching ability for many compounds in solution correlated with their pharmacological activity in *vivo* [419]. He supposed that their redox properties play the main role in both processes. The larger is the number of polar or redox groups in a given compound and the higher is the polarity of these groups, the higher is the efficiency of this compound as a quencher. Usually reducers and oxidizers are efficient quenchers of fluorescence [6, 57, 283, 284, 420]. On the other hand, the probability of changing the structure of an enzyme by sorption of a compound depends on polarity and charges of the compound.

For example, interaction of thirteen derivatives of the antiarrhythmic phenothiazine 10-acylaminopropionil with phospholipid membranes was studied in [414, 418]. Phenothiazines were the quenchers of probe fluorescence in liposomes. The binding constant was determined by fluorescence quenching of the probe, 3-methoxybenzanthron (Fig. 15.8).

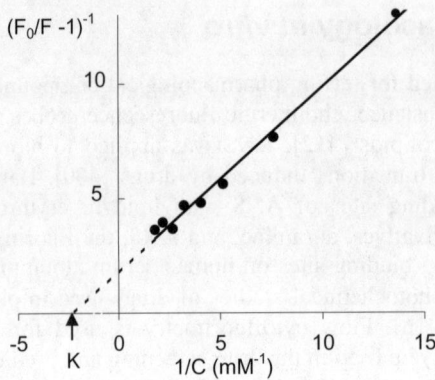

Fig. 15.8. Determination of the association constant (K) for the binding of ethmozin to phospholipid liposomes [418]. C is the concentration of ethmozin, F_0 and F are the fluorescence intensities of 3-methoxybenzanthron (5 μM) in suspension of liposomes (0.3 mg lipid/ml) without and with ethmozin, respectively

It was found that for 10 substances there is a correlation between the binding constant *in vitro* and the antiarrythmic activity on animals. For ethmozin derivatives the coefficient of correlation between the quenching constant and the activity was 0.86. The greater is the constant, the greater is the activity [414, 418]. However, three other substances characterized by the highest binding constants had low activities. It was assumed that a very high affinity to lipids can delocalize the drug in the organism and decrease its antiarrythmic activity.

15.7 Conclusion

The interaction of SecB *Escherichia coli* chaperone with model precursor protein, RBPTI, was studied using ANS. Binding of RBPTI by SecB-ANS resulted in an increase and blue shift in the ANS fluorescence. Two ANS emission centers exist in the proteins and complex: the short-lived center where ANS is strongly quenched by some polar aminoacid residues, and the long-lived center where ANS emits from the hydrophobic pocket. Estimation of the volumes and diameters of SecB, RBPTI and SecB-RBPTI using fluorescence life-time and polarization of ANS by a modified Levshin-Perrin equation was made. The effective diameter of SecB in aqueous solution is 33.4 Å. The diameter of the SecB-RBPTI complex equals to 49.8 Å. The volume ratio in the complex was equals to ~4 RBPTI per 1 tetramer SecB.

Not only spontaneous mobility of fluorescent probes, but also thermal and charge mobilities induced by a photon excitation take place in membranes.

Quenching of anthracene fluorescence by dimethylaminochalcone in liposomes is of a dynamic nature; probably, an exciplex is formed.

Fluorescent probes can be applied in pharmacological tests *in vitro*. The method was demonstrated with phenothiazines of various antiarrythmia activity.

16 Pyrene Monomers and Excimers in Membranes

16.1 Viscosity Measurements

The aromatic hydrocarbon pyrene is widely applied as a luminescence probe sensitive to the physico-chemical state of biological and model lipid membranes [19-21, 60, 369–372, 375, 382, 421–423]. Very long lifetime allows pyrene to be high sensitive to viscosity, temperature, etc. The pyrene "boom" was initiated by the simultaneously published reports by Vanderkooi & Callis [369] and Galla & Sackmann [370], where the ability of pyrene to form excimers was used as a basis for a new approach to the measurement of membrane viscosity. This extrapolation was borrowed from well known studies of the dependence of the rate of pyrene excimerization in organic solvents on their viscosity, which affects molecular diffusion (reviews [138, 151, 52–55, 57, 162]).

Excimerization process is described by the following scheme:

$$Pyr \xrightarrow{I_0} Pyr^* \rightarrow I_m + Pyr$$

$$Pyr \xrightarrow{I_0} Pyr^* \xrightarrow{Pyr} (PyrPyr)^* \rightarrow I_e + Pyr + Pyr$$

where Pyr is pyrene, Pyr^* is the excited pyrene monomer, $(Pyr\,Pyr)^*$ is the excimer, I_0, is the exciting light, I_m and I_e are the emissions of monomers and excimers, respectively.

In the measurements of the dependences of I_m and I_e on the concentration of pyrene and on other factors most researchers believed that it is the viscosity that determines the degree of excimerization, I_m/I_e.

However, such an approach proved to be inadequate. In [373, 424–427] the processes that influence pyrene excimerization in membranes were studied in detail. It was found that: (a) pyrene exhibits strongly luminesce not only in membranes but also in aqueous solutions; its monomer and dimer emission bands are similar to the monomer and excimer emission in membranes [424]; (b) at certain concentrations pyrene forms dimers, aggregates and clusters in water and in membranes [424, 427]; (c) in membranes of liver mitochondria at low concentrations, pyrene is not aggregated and does not form dimers, nevertheless, there are a few excimers; some pyrene molecules are adsorbed on proteins [373, 424]; (d) temerature variations do affect the quantum yields of both monomers and excimers; the temperature also influences redistribution of the probe between the lipid phase, proteins, and pyrene aggregates [424].

For pyrene dimers and clusters, the above scheme should be supplemented with:

$$(PyrPyr) \xrightarrow{I_0} (PyrPyr)^* \rightarrow Pyr + Pyr + I_d$$

where I_d is dimer emission.

16.2 Location and Diffusion of Pyrene

Some pyrene molecules in biomembranes are adsorbed on proteins, because a noticeable quenching of tryptophan fluorescence by pyrene is observed. For native and delipidized mitochondria the values of quenching of tryptophan fluorescence by pyrene differ insignificantly; at the same time, the excimer-monomer ratio F_e/F_m differs 5-fold (Table 16.1). This indicates that in mitochondria a considerable fraction of pyrene molecules is adsorbed on proteins even if the concentration of pyrene is low.

In membranes of sarcoplasmic reticulum pyrene, at low concentration, is mainly located in the lipid phase. At increased concentrations pyrene is adsorbed on Ca^{2+}-ATPase and quenches protein fluorescence. Fig. 16.1 shows the concentration curves of monomeric and excimeric luminescence of pyrene in sarcoplasmic reticulum membranes and lecithin liposomes [426]. It is seen that in the reticulum pyrene initially behaves as in liposomes.

From a certain concentration, its monomeric luminescence strongly increases due to binding by proteins (at these concentrations of pyrene the Ca^{2+}-ATPase fluorescence is being quenched).

Pyrene rapidly laterally moves in the region of the hydrocarbon tails of phospholipids in membranes [19, 369, 370, 421, 422]. However, it has been shown that pyrene is also capable of rapid transmembrane moving. Table 16.2 represents the values of quenching of pyrene luminescence in liposomal membranes by a number of compounds penetrating the membrane in different ways. It is seen that anthra-

Table 16.1. Limiting quenching of fluorescence of mitochondrial proteins by pyrene and the excimer-monomer ratio in native and delipidized mitochondria

Mitochondria	Quenching [%]	F_e/F_m
Native	27	0.47
Delipidized	23	0.10

Quenching was measured with excitation at 286 nm and emission at 335 nm. F_e/F_m was determined with excitation at 335 nm and emission at 393 nm (monomers) or 470 nm (excimers), the slits were 2 nm. Concentration of mitochondria is 0.6 mg protein / ml, pyrene concentration is 2.4 µM.

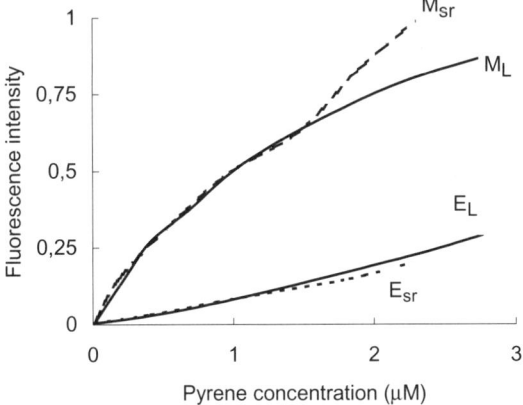

Fig. 16.1. Concentration curves for monomeric and excimeric fluorescence intensities of pyrene in membranes of sarcoplasmic reticulum (SR) and lecithin liposomes (L) [426]. M is monomer, E is excimer. Initial monomer τ in reticulum was 157 ns, in liposomes was 155 ns

cene and diS-C_3-(5) are capable of quenching both monomeric and excimeric bands even at low concentrations. ANS and phenol quench only the monomeric luminescence. Highly hydrophilic substances such as acrylamide or cesium have no effect even at high concentrations.

Thus, it can be assumed that the pyrene monomer has enough time to diffuse from the central part of the membrane to the polar regions where ANS and phenol

Table 16.2. Intensity of monomeric and excimeric fluorescence of pyrene in lecithin liposomes in the presence of various quenchers

Quencher	Concentration, [M]	Fluorescence intensity, [%]	
		F_m	F_e
Without quenchers	-	100	24.2
Anthracene	3.7×10^{-5}	35	9.7
Phenol	7.1×10^{-2}	67	24.2
ANS	1.5×10^{-5}	57	26.0
diS-C_3-(5)	0.5×10^{-5}	72	9.5
DL-tryptophan	1.5×10^{-4}	100	24.2
Acrylamide	1.2×10^{-1}	100	24.2
Cesium chloride	3.0×10^{-1}	100	24.2
Potassium nitrate	1.0×10^{-1}	100	24.2

15 mM Tris-HCl, pH 8.0. Concentration of pyrene is 15 μM, concentration of liposomes is 1 mg / ml. Excitation wavelength is 334 nm, the slits are 2 nm. The data are corrected for screening and reabsorption, as well for fluorescence of anthracene and ANS.

are incorporated [60], but it does not enter the aqueous phase. The excimer is probably unable to diffuse to the polar region of the membrane because it has twice the larger size and smaller lifetime.

The lifetime of pyrene monomer emission decreases with the addition of quenchers in all cases of quenching. The Stern-Volmer equation is valid for low values of quenching. The decrease in the lifetime of pyrene monomer emission could be the result of dynamic deactivation. The efficiency of quenching points to the affinity of the quenchers for the membrane and to their ability to accept vibronic excitation from pyrene.

Membrane viscosity calculated from F_e/F_m or from the concentrational self-quenching of monomeric emission was estimated by the most authors under aerobic conditions. However, pyrene is strongly quenched by molecular oxygen [138].

From all above said we might conclude that the measurements of membrane viscosity based on pyrene luminescence were not correct. A similar conclusion has been independently made by Blackwell et al. [427]. Probably, more correct approach is using the phospholipid derivatives of pyrene [376, 377]. These probes in lipid membranes at temperatures above phase transition show a two-dimensional diffusion-controlled kinetics in excimer formation [376]. Quenching of the fluorescence of pyrene-labeled phospholipids by dibromolipids was used in [377] to determine the chain length dependence of the bilayer depths of the pyrenyl moieties. 1-palmitoyl-2-(pyrenyl-n-acyl)-phosphatidylcholines with end-labeled pyrenyl chains varying in length, n, from 4 to 14 carbons were examined. It was concluded that these pyrene derivatives mimic the natural phospholipids.

16.3 Detection of Oxygen by Pyrene Emission

Pyrene emission in membranes is strongly quenched by oxygen (Table 16.3 and Fig. 16.2). Adding succinate to mitochondria results in the consumption of oxygen due to the oxidation of succinate in the respiratory chain, that is accompanied by an increase in the lifetime and quantum yield of pyrene. With the depletion of oxygen a maximum luminescence intensity is reached which is invariant with time

Table 16.3. Pyrene fluorescence in mitochondrial membranes under aerobic and anaerobic conditions

Oxygen concentration	F_m	F_e	F_e/F_m	τ_m, [ns]
130 µM	95	11.5	0.12	105
~1 µM	150	23.0	0.15	157

Incubation medium contained 10 mM Tris-HCl, 10 mM KH$_2$PO$_4$, 50 mM KCl, 150 mM sucrose, pH 7.5. Mitochondrial concentration is 0.8 mg protein/ml. Pyrene concentration is 10 µM. Excitation wavelength is 336 nm, emission wavelength is 393 nm (monomers) and 480 nm (excimers) Monochromator slits are 2 nm

(Fig. 16.2). The quenching constant, K_q, could be calculated from the Stern-Volmer equation:

$$K_a = \frac{\tau^{-1} - \tau_0^{-1}}{[Q]} \qquad (16.1)$$

where τ_0 and τ are the lifetimes in the absence and in the presence of the quencher, [Q] is the concentration of the quencher. The value of K_q calculated from the data in Table 16.3 is 2.4×10^{10} M^{-1}s^{-1}. Such high quenching constant is due to a considerably higher actual concentration of oxygen in the membrane than in the aqueous phase, because the solubility of oxygen in organic solvents is usually 5- to 20-fold greater than in water. The ratio of oxygen amount between lipid and aqueous phases is about 4.4 [371].

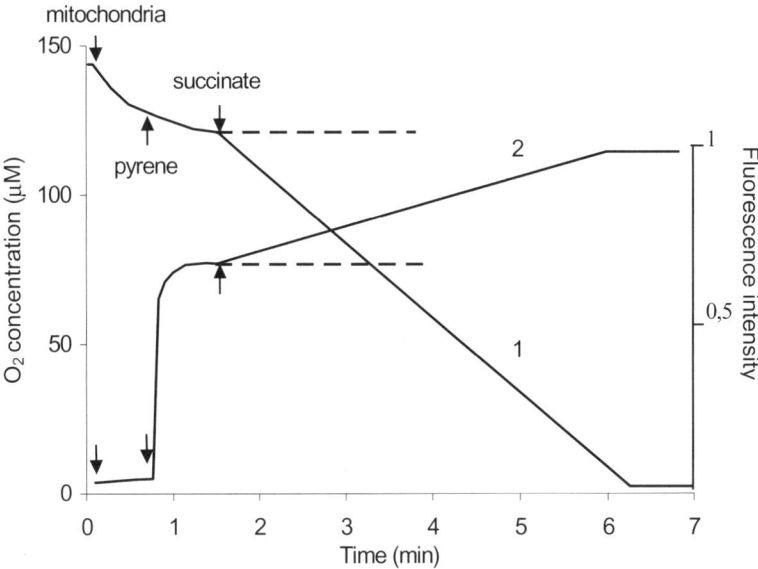

Fig. 16.2. Oxygen uptake (1) and the increase in the fluorescence intensity of pyrene monomers (2) in mitochondria placed in a hermetically sealed cuvette [425]. The incubation medium is the same as in Table 16.3. Excitation at 336 nm, emission at 393 nm, slits arc 2 nm. Concentrations: mitochondria is 0.8 mg protein / ml, pyrene is 1 µM, succinate is 10 mM. The dashed lines are oxygen concentration and fluorescence intensity in the absence of succinate

Under anaerobic conditions the pyrene monomer in mitochondria has lifetime of 157 ns (Table 16.3). In oxygen-free organic solvents the lifetime of pyrene is essentially higher. For instance, in cyclohexane $\tau = 370$ ns [138]. In mitochondria pyrene is quenched not only by oxygen, but also by phospholipids, that agrees with [371].

16.4 Vibronic Peaks as Indicators of Membrane Polarity

The vibronic bands of the monomeric emission of pyrene are known to be sensitive to the polarity of the medium [261, 262]. Fig. 16.3 shows the positions of the vibronic bands. It can be seen from this Figure that there is a population of pyrene molecules not involved in excimerization. These molecules are quenched by phenol.

Table 16.4. Intensity of vibronic peaks of pyrene luminescence in various environments [373]

	I	II	III	IV	V
Liposomes-a)	1.0	0.71	1.04	1.15	1.23
Liposomes-b)	1.0	0.67	0.92	0.85	0.98
Native mitochondria	1.0	0.68	0.92	0.83	0.95
Delipidized mitochondria	1.0	0.69	0.92	0.83	0.94
Bovine serum albumin	1.0	0.73	0.87	-	0.98
Water	1.0	0.55	0.56	-	0.88
Ethanol	1.0	0.68	0.87	-	1.01
n-Hexane	1.0	1.01	1.49	1.15	1.18
Hydrocarbon solvents [185]	1.0	1.08	1.74	1.3	1.4
Aromatic solvents [185]	1.0	0.7	0.9	0.85	0.97
Gas phase [184]	1.0	-	2.43	-	-

Excitation wavelength is 335 nm; slits are 2 nm. a) pyrene / lipid ratio is 1:600; b) pyrene / lipid ratio is 1:100, F_e/F_m=0.24.

Table 16.4 summarizes the data on the intensities of the vibronic bands of pyrene in various environments. These data suggest that the luminescence of pyrene monomer in the membrane originates mainly from the polar lipid phase, i.e. from the region of the polar heads of phospholipids rather than from the hydrocarbon region of the lipid layer. During the lifetime, the monomers seem to be able to diffuse to the polar region, where they emit photons. The monomers remaining in the hydrocarbon region are excimerized. In fact, the vibronic bands of pyrene in lecithin liposomes and in mitochondria are similar to those of pyrene in ethanol and albumin rather than in n-hexane (Table 16.4).

The ratio of the intensities of vibronic bands in membranes depends on the concentration of pyrene. This can be due to both the changes in the location of pyrene in the membrane and the perturbation of the lipid bilayer caused by high concentrations of pyrene.

Our data on the sensitivity of vibronic bands of pyrene in membranes to the microenvironment are used in studies of various biological objects; for instance, in zoospores [382]. A similar approach was used for 16-(1-pyrenyl)-hexadecanoic acid in liposomes [374]. The dependence of pyrene vibronic intensities on the NaCl concentration was measured in [428].

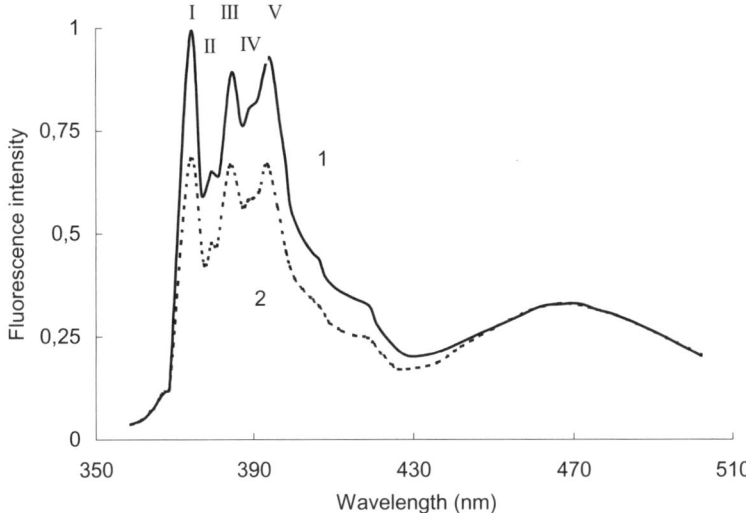

Fig. 16.3. Fluorescence of pyrene monomers and excimers in mitochondria before (1) and after (2) the addition of 100 mM phenol [373]. Mitochondrial concentration was 0.17 mg protein/ml. Excitation wavelength was 335 nm, slits were 2 nm. I, II, III, IV, V are the vibronic peaks

16.5 Conclusion

The conventional measurements of membrane viscosity based on pyrene luminescence in biomembranes were, as a rule, incorrect. In a membrane the pyrene monomer moves not only laterally but also across the membrane. In the lipid membrane the pyrene monomer emits from two sites: the polar region and the subpolar region rather than the region of hydrocarbon tails of phospholipids. In biomembranes a considerable fraction of pyrene molecules is bound to proteins. Under aerobic conditions pyrene luminescence is strongly quenched by oxygen. A high sensitivity of pyrene luminescence to oxygen can be used as the basis for the measurement of oxygen uptake in biological suspensions.

17 Photomodulation of Enzyme Activity

Modulation of the enzyme activity by light is an important task from the fundamental and applied points of view. Light stimulation is widely used in cellular biology and medicine. Absorption of a photon by an enzyme molecule may induce conformational changes and relaxational dynamics, which result from vibrations and rotations during nonradiative transitions. Also, a local "heating" of the enzyme globule in the sub-nanosecond time scale may occur.

Enzymes may be directly or indirectly photomodulated [302]. Direct photomodulation is a photo-production of the reaction product, photo-dissociation of an inhibitor, photochemistry of amino acids, etc. Indirect photo-modulation is gene expression and stimulation of non-enzyme-containing photoreceptors. A theoretical model on the light activation of redox-enzymes was suggested in [524]. A lot of experimental data on activation and inactivation by UV or visible light of various biochemical reactions of the enzymes isolated from animals and plants are available. Let us consider some examples.

17.1 Photoactivation of Enzymes

Light stimulates the activity of cyclooxyganese, protein kinase, ornithine decarboxylase of skin, DNA photolyase, etc. (review [429]).

Reversible photoregulation of activities of various proteins (papain, concanavalin A, and chymotrypsin) was achieved in [430] by two approaches. The first approach involves a covalent attachment of photoisomerizable components as azobenzene units or thiophenefulgide components. In one of the photo-isomer states of the photoactive units, the tertiary structure of the proteins is retained and their activity is switched on. In the complementary photoisomer state, the structure is distorted and the activity is switched off. The second approach involves immobilization of proteins in photoisomerizable polymers. In this system photostimtilated transport of the protein across the polymer matrices triggered on-off the functions of the encapsulated protein.

Electron flow through the molybdopterin cofactor to its acceptor nitrate is involved in the flavin-mediated signal transduction for condiation in light [431]. In [534] a photoinduced electron transfer in glutathion reductase was observed.

The folate chromophore in the *E. coli* DNA photolyase serves as an antenna, transferring light energy to the reduced flavin ($FADH_2$) reaction center [432]. The fluorescence of enzyme-bound folate is quenched upon flavin binding. Energy transfer from folate to $FADH_2$ is sensitive to the stereo-chemical configuration. The efficiency of energy transfer is directly proportional to the fluorescence quantum yield of folate.

A hypothetical photobiological role of dityrosine in the outer layer of intact yeast ascospores and the protection of the spore genome from mutagenic UV light were discussed in [433].

In [434] quantum yield and spectral action data on thymine dimer repair were obtained under aerobic conditions. Dimer repair was initiated either by direct excitation of $FADH_2$ or by the excitation of the enzyme (pterin) followed by singlet-singlet energy transfer to $FADH_2$. The quantum yield observed for dimer repair with native enzyme (=0.72) was similar to that observed with the enzyme containing only $FADH_2$.

Photolyase can repair DNA pyrimidine dimers by converting the energy of light to chemical energy by means of a specific tryptophan residue of this enzyme. In this case tryptophan acts as an electron donor to the site of the DNA damage (cit. by [247]).

Oligopyrimidines covalently linked to ellipticine derivatives form duplex and triplex structures with target single-stranded oligopurine sequences. They also bind to duplex DNA at homopurine-homopyrimidine sequences where they form local triplet helices. Irradiation by wavelengths, longer than 300 nm, of an oligonucleotide-ellipticine conjugate induced a) cleavage of the target at the bases located in close proximity to the dye and b) cross-linking of the target sequence to the derivatized oligonucleoudes [435].

In [436] it was shown that photo-activation of nitrile hydratase resulted in oxidation of iron in the enzyme. Tryptophan residues in the enzyme induced the oxidation of the iron via an energy transfer process. Nanosecond flush photolysis revealed that this photo-activation process was completed within 50 ns. In [213] a conformational change accompanying the phototransformation of phytochrome-A was observed using fluorescence quenching. Changes in the protein secondary structure seem to be confined to the N-terminal region, where a photoreversible increase in the α-helix content occurs. In [437] an un-relaxed oat phytochrome intermediate was detected by time-resolved circular dichroism studies.

17.2 Photodesorption

Photostimulated desorption is observed as a result of vibrational excitation of different chemical systems [438]. Several kJ/mol (the energy of vibronical quanta) are sufficient to desorb substances from a surface. For instance, desorption of NH_3 and other small molecules from a solid surface irradiated by an IR laser was observed [439]. The energy of vibrational excitation was transferred from the "support" to NH_3 that led to the breakage of the bond and desorption. Desorption of a short dye-labeled DNA from a silianized glass surface under 514-nm laser beam was observed in [310].

Photoexcitation is well known to be able to produce dissociation of heme-bound ligands such as CO and O_2 [57, 397]. The energy of binding of CO to heme is about 28 kJ/mol [440]. Illumination of CO-inhibited mitochondria leads to desorption of CO from cytochrome oxidase and repair of mitochondrial respiration [441]. In [442, 443] hot vibrations in photoexcited hemoglobins were observed using picosecond Raman spectroscopy.

Coherently transport of NH stretching quanta along a chain of hydrogen-bonded peptide groups was demonstrated by simulating the α-helical peptide poly(L-alanine) [444]. The dependences of vibrational energy motion on time were obtained.

17.3 Photochemical Processes in Alcohol Dehydrogenase

One of the most suitable enzymes to study the processes of the transformation of photoexcitation energy is horse liver alcohol dehydrogenase. The structure of enzyme-NAD^+ and enzyme-NADH complexes is known due to X-ray analysis [327, 445, 446]. NADH is able to fluoresce, its fluorescence intensity increases and the spectrum shifts to shorter wavelengths when it binds to the enzyme [70, 447]. Unlike NAD^+, NADH has a sufficiently wide absorption band at 340 nm, and therefore, it can be easily spectrophotometrically detected [60, 70, 448].

When NADH is bound to the enzyme, it quenches the fluorescence of Trp-314 via electronic excitation energy transfer [60, 165, 447] (Chapter 13), i.e. it is possible to selectively excite NADH bound to the protein via Trp-314. NADH bound to the enzyme has a long lifetime (many times longer than the lifetime of a free NADH in water), ~1.8 ns [60]. Varying the pH or the concentration of the substrate can easily shift the equilibrium to any side:

$$C_2H_5OH + NAD^+ \rightarrow C_2H_4O + NADH + H^+$$

Fig. 17.1 shows the absorption spectra of alcohol dehydrogenase, NAD^+ and the alcohol dehydrogenase-NADH complex; the latter being formed upon reduction of NAD^+ by ethanol. It can be seen that the enzyme-NADH complex possesses a high photochemical stability. When this complex is illuminated with a 450 W xenon lamp for 1.5 min, the protein and NADH are not destructed.

Fig. 17.2 shows the fluorescence intensity (it is called a "fluorescence photoresponse") of NADH illuminated through the filters UVL-6 (transmission from 320 nm to 380 nm) and UVL-1 (transmission from 250 nm to 380 nm) under different experimental conditions. The experiments were carried out with the device "SLM-4800" without the excitation monochromator; the filters were used in combination with a heat filter. A solution of the enzyme (2 ml) was thoroughly thermostated. In addition to the monochromator in the emission channel a filter SZS-22 or SZS-19 was used to reduce the scattered excitation light. The slits on the emission monochromator were set to 4 nm. The wavelength of fluorescence was 450 nm.

The light in the wavelength range between 320 and 380 nm (transmitted by UVL-6 filter) induces a slow decrease in the intensity of NADH in the presence of

17.3 Photochemical Processes in Alcohol Dehydrogenase

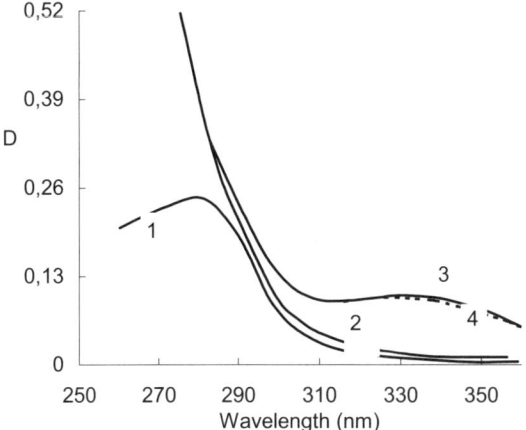

Fig. 17.1 Absorption spectra of alcohol dehydrogenase (1), NAD^+ (2) and the NADH-alcohol dehydrogenase complex (3 and 4) [449]. The complex was obtained by reducing NAD^+ by ethanol. Spectrum 4 is measured after illuminating the complex (through a heat filter) with the 450-W xenon lamp for 1.5 min

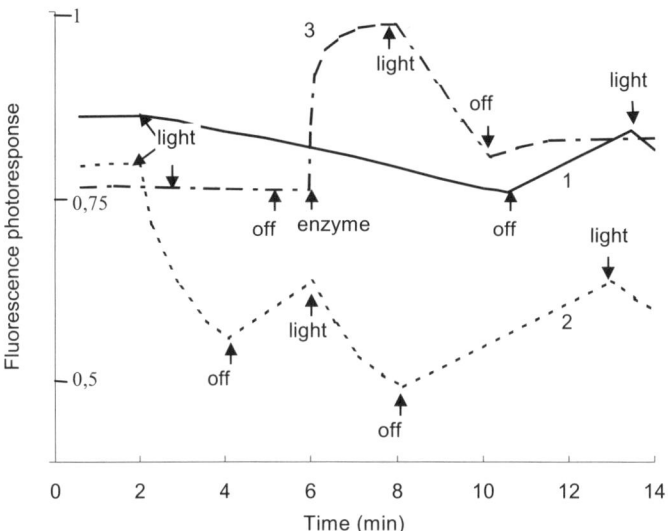

Fig. 17.2. Fluorescence photo-response of NADH in the presence of alcohol dehydrogenase under UV-illumination [449]. 1) a solution of NADH with the enzyme illuminated through UVL-6 filter; 2) the same solution illuminated through UVL-1 filter; 3) a solution of NADH illuminated through UVL-1 filter. Concentrations are: NADH is 10 - 26 µM, enzyme is 3 µM; the buffer is 12 mM Tris-HCl, pH 7.5

the enzyme. In the absence of the enzyme no changes are observed. When the light is switched off, the fluorescence intensity of NADH in the presence of enzyme slowly increases although the fluorescence intensity does not reach the initial level. With the subsequent repeated switching on-of the illumination fluorescence decreases again. A similar phenomenon is also observed when the filter UVL-1 is used (transmission between 250 and 380 nm); the effect being even more pronounced (Fig. 17.2, Table 17.1). The considerable increase in the amplitude of photo-response with the filter UVL-1 as compared to UVL-6 is caused by two factors: a greater total intensity of the excitation light and the presence of energy transfer from Trp-314 to NADH.

It is not important whether the NADH-enzyme complex is formed by adding NADH or by reducing NAD^+ by ethanol (Fig. 17.2 and Table 17.1). However, the amplitude of the photo-response is greater in the latter case. Probably, in the former case many NADH molecules remains in solution, since NADH is in more amounts than the enzyme. In fact, the used amount of the enzyme can bind only 6 µM NADH out of the 26 µM of the added NADH (two NADH molecules per enzyme molecule). Therefore, about 20 µM of NADH remains in the solution and hinders the measurement, although the lifetime and the quantum yield of NADH bound to the enzyme are about 3 times greater than those of free NADH in solution [60, 165, 447]. In the second case, the amount of NADH produced in the reaction was no more than 10 µM, and about 6 µM were bound to the enzyme. Hence, the fluorescence was mainly due to the NADH molecules bound to the protein.

According to the X-ray structural data [327, 445, 446], an NADH molecule bound to alcohol dehydrogenase is "buried" in a hydrophobic "pocket" so that the

Table 17.1. Changes in the fluorescence intensity of NADH illuminated by the UV in the presence and in the absence of the enzyme under different experimental conditions [449]

Time (min)	Light (switch on/off)	Fluorescence intensity in the solution:			
		NADH	NADH Enzyme	NAD^+ Enzyme C_2H_5OH	NADH BSA
1	Off	0.71	0.93	0.80	0.85
2	Off	0.71	0.93	0.80	0.85
3	On	0.69	0.82	0.63	0.79
4	On	0.69	0.72	0.55	0.79
5	Off	0.69	0.81	0.59	0.80
6	Off	0.69	0.90	0.64	0.81
7	On	0.67	0.77	0.53	0.77
8	On	0.67	-	0.48	-

Illumination through the filter UVL-1. Concentrations are: NADH of 26 µM, enzyme of 3 µM, NAD^+ of 75 µM, ethanol of 30 mM, bovine serum albumin (BSA) of 3 µM. The solution that contained NAD^+, ethanol and the enzyme was preequilibrated; the enzyme-NADH complex was formed at a concentration of about 3 µM. The buffer was 12 mM Tris-HCl, pH 7.5.

pyridine ring of NADH is bound to the protein and it inaccessible to the solvent. The vibronic quanta released in the course of the vibrational relaxation of the excited NADH first penetrate the protein globule and then pass to the aqueous solution. The difference between the energies of the long-wavelength electronic transitions related to the absorption at 340 nm (29400 cm^{-1}) and the emission at 450 nm (22200 cm^{-1}) corresponds to the energy of vibrational relaxation in the NADH molecule (7200 cm^{-1}=80 kJ/mol). This energy is sufficient to break the bond between the enzyme and NADH. Therefore, during intense illumination of alcohol dehydrogenase, NADH could desorb from the enzyme.

Along with photodesorption, other reasons could be responsible for the decrease in NADH fluorescence. They are (a) photodestruction of NADH; (b) photooxidation of NADH; (c) a transition of NADH to a long-living state, in particular, the triplet state; d) trivial heating of a solution. Let us consider listed possibilities. (a) Though photodestruction of NADH takes place, but the rate of this process is hundred times slower than that of the fluorescence photo-response (compare spectra 3 and 4 in Fig. 17.1 with Fig. 17.2). (b) During powerful illumination NADH is able to transform into metastable state, possesses an increased reaction ability [450]. About photooxidation of NADH by flavins see Chapter 18. Photooxidation of NADH by any electron- acceptor groups could be possible. However, alcohol dehydrogenase is no contains any electron acceptor groups. The facts that the fluorescence photo-response is partially reversible as well as the absence of changes in the absorbance (Fig. 17.1) contradict the assumption about photooxidation (Table 17.1). (c) The lifetime of the triplet state of NADH is 140 ms at low temperature [365]. Hence, at room temperature and, what is more important, under aerobic conditions the photo-response due to NADH transition to

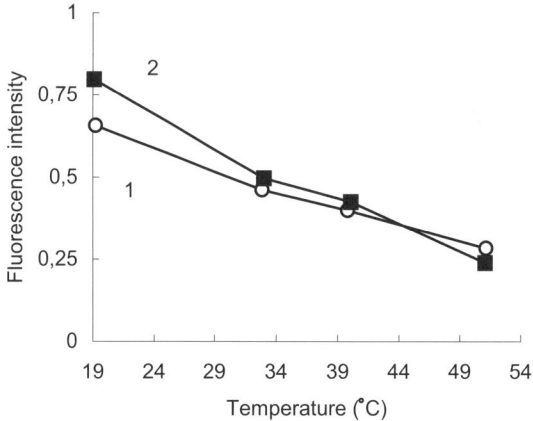

Fig. 17.3. Temperature quenching NADH fluorescence in the presence and in the absence of the enzyme [449]. 1) NADH in a buffer; 2) NADH in the presence of alcohol dehydrogenase. Buffer is 12 mM Tris-HCl, pH 7.5

the triplet state should be neglected. (d) Well known that a temperature decreases the fluorescence intensity. Fig. 17.3 shows the effect of temperature on fluorescence of NADH in aqueous solution in the absence and in the presence of the enzyme. In these experiments, no intense illumination was used: the fluorescence was measured with very low intensity of excitation (with a monochromator in the excitation channel; the slits were 2 nm). To decrease the fluorescence two-fold, the solution should be heated up to ~50°C. Heating leads to a decrease in the fluorescence of not only NADH-enzyme complex, but also of free NADH. Powerful illumination of the enzyme-NADH complex (under thermostated conditions) increased the temperature not more than 0.5°C; it was experimentally proved. The illumination of NADH solution, without enzyme, produced no significant change in fluorescence (Table 17.1).

An instantaneous "local heating" of the photoexcited complex occurs, hence, NADH can be desorbed from a protein. The presence of the NADH photodesorption is supported by the results obtained with the system NADH/albumin (Table 17.1) Qualitatively, the system behaves in the same way as the system NADH/alcohol dehydrogenase. It is known that bovine serum albumin binds various compounds, including NADH. During the illumination of NADH bound to albumin direct excitation and sensitization of NADH via tryptophans are possible. The amplitude of the photo-response in the NADH-albumin system is much lower

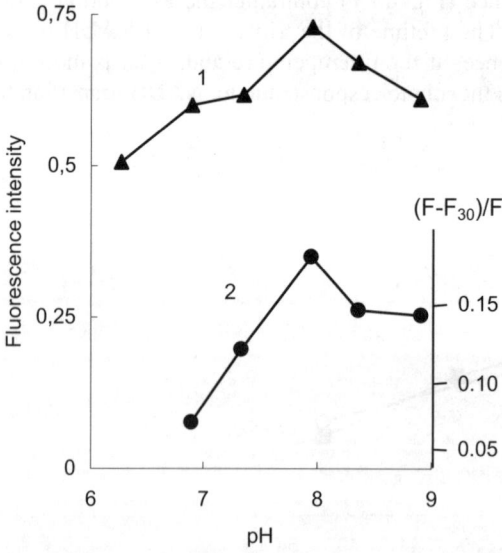

Fig. 17.4. pH dependences of the "dark" fluorescence of the NADH-alcohol dehydrogenase system (1) and of the fluorescence photoresponse of the same system under illumination with UV- light (2). Medium: 30 mM Tris-HCl, 30 mM KH_2PO_4. Concentrations are: NADH is 10 μM, the enzyme 3 μM. F is the initial intensity; F_{30} is the intensity after 30 s illumination [449]

than in the NADH-enzyme complex. This suggests a specificity of the NADH-enzyme complex.

Thus, all the data obtained make us to conclude that the NADH-enzyme system demonstrates a specific fluorescence photo-response under the beam of the UV-light. Probably, photodesorption of NADH from the protein is mainly due to the instantaneous "local heating" of the complex during vibrational relaxation and internal conversion of the excited NADH.

Fig.17.4 shows the pH dependence of the fluorescence intensity of the NADH-enzyme system and the rate of fluorescent photo-response in the course of illumination of the system. The maximum fluorescence intensity in the dark (without intense illumination) is observed at pH=8. The intensity decreases of acidic and alkaline pH. The maximum rate of fluorescence photo-response to illumination is also at pH 8. These data point to the specificity of the photo-response, because alcohol dehydrogenase has optimum enzyme activity at pH=8 [228]: the enzyme activity in the forward and backward directions reaches an equilibrium at this pH. The photo-response peak is detected at the same pH.

When the chromophore is "buried" inside the globule, the enzyme under illumination could be also photodenaturated. When NADH bound to the enzyme is excited (directly or via Trp-314), a photoconformational relaxation of the protein structure occurs, and, possibly, it can result in denaturation. This assumption is proved by the fact that after illumination the fluorescence of NADH in the dark increases but does not reach the initial level (Table 17.1).

17.4 Photolysis of Flavin in NADH Dehydrogenase

Complex I (NADH dehydrogenase) is the key enzyme of the respiratory electron transport chain. This complex catalyzes ubiquinone reduction and transmembrane electron transfer coupled to ATP synthesis. The NADH dehydrogenase complex of bovine heart mitochondria was shown to consist of 42 subunits and contain FMN and six iron-sulfur clusters, one of them (Fe_4S_4) being located near FMN.

In addition to ubiquinone, electrons from NADH can be transported through the enzyme complex to other acceptors. The catalytic properties of NADH dehydrogenase in submitochondrial particles and isolated detergent preparations are usually studied using artificial electron acceptors. The enzymatic oxidation of NADH mediated by acceptors such as ferricyanide circumvents ubiquinone. Ferricyanide reduction on high-potential iron-sulfur clusters occurs at a site different from the site of ubiquinone reduction. On the other hand, acceptors such as para-nitrotetrazolium violet (p-NTV) and dichlorophenolindophenol (DCPIP) are thought to be reduced by ubiquinone.

The sequence of the electron transfer through the redox centers of the NADH dehydrogenase complex is still insufficiently understood. It is widely believed that electron transfer from NADH to ferricyanide is mediated by flavin and iron-sulfur clusters, i.e. that all redox components of NADH dehydrogenase (flavin, iron-sulfur centers, and ubiquinone) are involved in the rapid relay passing of an elec-

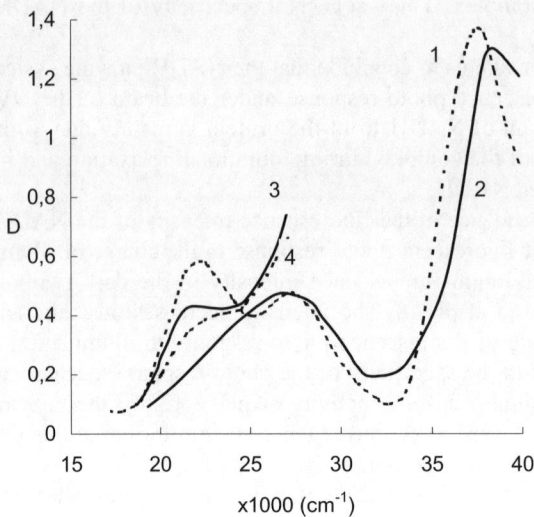

Fig. 17.5. The absorption spectrum of FMN in water: *(1)* before illumination; *(2)* after illumination; *(3)* in the presence of 280 μM $FeCl_3$ before illumination; and *(4)* in the presence of 280 μM $FeCl_3$ after illumination. Exposure time of illumination was 12 min

tron to *p*-NTV or DCPIP. These schemes of electron transfer are based on the redox potential levels of the electron carriers. However, the redox potential difference itself cannot be regarded as evidence for the given electron transfer pathway. Moreover, it was shown that the rate of ferricyanide-dependent oxidation of NADH by one of the enzyme fractions depleted with flavin (but enriched with iron) is many times higher than the rate of the reaction catalyzed by the fraction enriched with flavin.

The goal of our work [535] was to study the role of flavin and ubiquinone in the enzymatic oxidation of NADH activated by ferricyanide, *p*-NTV, DCPIP, and membrane-bound ubiquinone.

The fraction of 0.2-μm protomitochondrial particles containing intact NADH dehydrogenase was isolated from a suspension of bovine heart mitochondria. No detergents were used to keep the respiratory chain intact and to avoid the detergent-induced effect on the enzyme interaction with acceptors. Optical transparency of the protomitochondrial suspension is a methodological advantage of the use of this fraction.

To measure the NADH dehydrogenase activity, a suspension of the protomitochondrial fraction in 5 or 20 mM Tris or phosphate buffer (pH 7.5; 20°C) was poured into a 0.5- or 1-cm quartz cuvette of a Specord UV Vis or M-40 spectrophotometer. Oxidation of NADH was measured as absorption changes of NADH (at 340 or 360 nm) or electron acceptors: ferricyanide (at 410 nm), DCPIP (at 600 nm), and formazan, a product of p-NTV reduction (at 530 nm).

Photochemical reactions were carried out in a thermostatically controlled (tem-

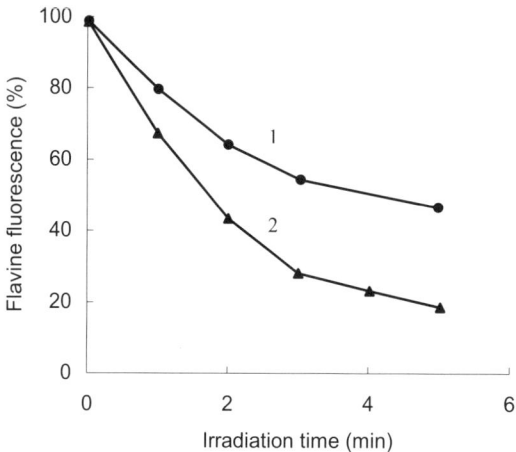

Fig. 17.6. Photobleaching of *(1)* free FMN and *(2)* protomitochondrial FMN during 5-min illumination. The degree of photobleaching was measured immediately after exposure as a decrease in the flavin fluorescence intensity. Flavin fluorescence was excited in the photobleached band at 450 nm and detected at 525 nm

perature, 14°C) metal holder equipped with a frontal window and a back reflector. Protomitochonrial flavins (in the experiment) or free FMN (in the control) were illuminated for 1–60 min with a 436-nm emission line of an SVD-120A lamp through an SZS-8 glass filter and a heat filter. Samples of small volumes for measurements of enzymatic activity and FMN photobleaching were collected during illumination.

Exposure of free FMN to intense blue light causes its irreversible destruction and, hence, disappearance of the absorption band at 450 nm. This is also accompanied by significant changes in the UV absorption region (Fig. 17.5). Photobleaching of FMN is accompanied by fluorescence disappearance (Fig. 17.6). This fact can be used for assaying NADH dehydrogenase, because it is the only membrane-bound enzyme containing FMN, a high-activity flavin species (other enzymes contain FAD). Short-term (1-2 min) exposure of mitochondria to blue light is in some cases accompanied by flavin-dependent oxygen uptake and ATP synthesis (see next chapters). NADH dehydrogenase is thought to be a possible mediator of this process. The flavin moiety of NADH dehydrogenase is the main contributor to the total flavin fluorescence of bovine heart mitochondria. The rate of FMN photobleaching in protomitochondrial NADH dehydrogenase is significantly less than the rate of photobleaching of free FMN (Fig. 17.6). Therefore, there is a certain group near the FMN binding site in protein that is able to deactivate effectively the excited state of FMN, thereby preventing FMN from photobleaching. Probably, the nearby Fe_4S_4 cluster plays the role of such a group. The *d*-orbital electrons of the iron atoms of the cluster are at a relatively large distance from the core of the cluster (5-10 Å). Therefore, iron interacts with flavin at this distance between the centers. This conclusion is supported by the fact that FMN in solution produces complexes with ferric ions (with the corresponding changes of the absorption spectrum). The photobleaching efficiency of this complex is reduced manyfold (Fig.17.6).

A fraction of protomitochondrial flavin does not undergo total photobleaching: after 5–10 min of exposure; the curve reaches a plateau (Fig. 17.6). This is due to the fact that, in addition to NADH dehydrogenase, FAD-containing flavoproteins also contribute to the flavin fluorescence of protomitochondria.

Photobleaching of protomitochondrial FMN was not accompanied by any changes in the tryptophan emission (not shown), which suggests an absence of protein conformation changes. However, photobleaching caused a decrease in the NADH oxidation rate in the respiratory chain (Fig. 17.7). These facts directly indicate that FMN is an essential component of the NADH: ubiquinone reductase activity. It should also be noted that the degree of the activity change differs substantially from the degree of photobleaching. For example, a 5-min exposure to blue light caused a 50% decrease in the flavin fluorescence yield, but only a 30% decrease in the NADH: ubiquinone reductase activity. Probably, there is a pool of FMN molecules in protomitochondria that do not belong to active NADH dehydrogenase. Perhaps, this pool belongs to "silent" enzyme molecules.

Photobleaching had no effect on the ferricyanide-activated oxidation of NADH (Fig. 17.7). Therefore, it is hardly probable that FMN is an intermediate of the reaction of electron transfer from NADH to ferricyanide. Similar results were obtained with p-NTV and DCPIP (Fig. 17.7).

Moreover, the addition of these acceptors in the absence of NADH and without illumination had no effect on the flavin fluorescence of the system. Therefore, the FMN binding site is at a considerable distance from the binding sites of the acceptors. As a result, reduced FMN is unable to donate electrons to these acceptors.

Although dithiothreitol significantly inhibits ferricyanide-dependent oxidation of NADH, it had no effect on the flavin fluorescence of the system studied in our experiments. This also can be regarded as evidence that flavin is not involved in the ferricyanide-dependent enzymatic oxidation of NADH.

In contrast to NADH oxidation in the respiratory chain, the NADH: ferricyanide reductase reaction is insensitive to rotenone, a specific inhibitor of electron transport in the native mitochondrial NADH dehydrogenase. The insensitivity to rotenone is due to the fact that ubiquinone is by no means involved in the NADH: ferricyanide reductase reaction. Similar results were obtained with p-NTV and DCPIP.

Thus, it may be concluded that the NADH dehydrogenase complex is able to catalyze two types of reactions: (1) reduction of ubiquinone or its analogs in the reaction mediated by flavin and iron-sulfur clusters and (2) reduction of artificial (and, possibly, some natural) acceptors in the reaction mediated by iron-sulfur clusters, rather than flavin and ubiquinone. The former reaction is rotenone-sensitive, whereas the latter is rotenone-insensitive.

In [536] the globule diameter and some properties of NADH dehydrogenase was estimated by time-resolved technics and correlated confocal microscopy.

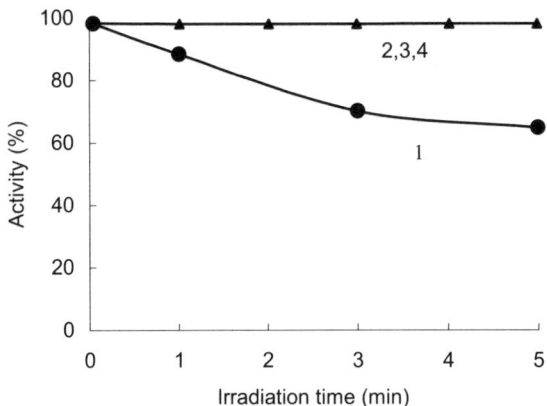

Fig. 17.7. Activity of protomitochondrial NADH dehydrogenase after illumination: *(1)* the NADH: ubiquinone reductase activity; *(2)* the NADH: ferricyanide reductase activity, *(3)* the NADH: p-NTV reductase activity; and *(4)* the NADH:DCPIP reductase activity. The NADH dehydrogenase activity was measured by changes in the optical density of NADH or electron acceptors and expressed as the percentage of the maximum level

17.5 Conclusion

The data point to the possibility of controlling the "dark" enzymatic processes by light.

Fluorescence analysis showed a high sensitivity of the NADH/alcohol dehydrogenase complex to the UV illumination. A rapid decrease in the fluorescence intensity of NADH (photo-response) is observed under the UV beam. No decrease was found in the solution of NADH without the protein. The maximum rate of photo-response was observed at pH 8.0, i.e. at the biochemical optimum for the enzyme. Probably, the main reason for the fluorescence photo-response is the desorption of NADH from the protein during the internal conversion and vibrational and conformational relaxations after photoexcitation of Trp-314 and NADH.

Photobleaching of FMN in the NADH dehydrogenase complex in protomitochondrial particles was found. This photobleaching prevents the NADH:ubiquinone reductase activity, but does not influence on the NADH oxidation in the presence of artificial electron acceptors.

18 Photoactivation of Animal Membranes and Their Chromophores

18.1 Photoinduced Membrane Activity

The study of the transformation of photoexcitation in membranes from animal cells is interesting from the viewpoint of the establishment of key differences between these membranes and photosynthetic membranes as well as principle common features of both kinds of membranes. Moreover, biochemical reactions in animal cells run in incomplete darkness: the phenomena of chemiluminescence and bioluminescence are well known [25, 29–34, 451]. The appearance of excited states has been found in some "dark" animal enzymes [451–454]. The functional role of light emission for animal cells remains unclear. It can be emphasized that the absence of the emission is not evidence for the absence of electronically excited states, because excitation may be transferred and stored without emission.

In recent years the interest in the effect of optical radiation on animal membranes has greatly increased [455]. Intense illumination of biopolymers and membranes usually results in their damage: structural changes and the loss of function [60]. For example, in [456] isolated rat liver mitochondria were incubated in the presence of photofrin and irradiated with the wavelength of 365 nm. After 45 s of irradiation (30 W/m^2) the coupling, defined as stimulation of respiration by externally added ADP, was totally lost; in contrast, the membrane potential created by addition of succinate or ATP was only slightly affected [456].

However, sometimes activation of specific processes is observed. In particular, the light-induced activation of bioelectric impulses in nerve cell was shown [457]. Illumination of mitochondrial enzymes complexes incubated with flavins and EDTA resulted in reduction of cytochromes [458]. Illumination of an isolated cyanide-inhibited cytochrome *c* oxydase by visible light induces an oxygen uptake [459]. This oxygen uptake was interpreted as increasing in the electron-transfer enzyme activity. When bacteriorhodopsin incorporated in phospholipid liposomes in the presence of mitochondrial ATPase is illuminated, ATP is synthesized [460, 461]. Rhodopsin is not activated in rods and cones without light [20, 38]. Photosynthesis in plants and bacteria is an important field of investigations [30-38]. In algae treated with white light, 15-ns dye-laser pulses (620-690 nm) induced chloroplast movement [462]. This movement depends on light polarization relative to the cell axis. It was shown that human lymphocytes are sensitive to the radiation of a He-Ne laser |463], etc.

Photolysis of the biologically inactive ATP precursor, "caged ATP", was used for the stimulation of tension of muscle fibre due to an ATP-actomyosin interaction. Caged ATP is a chemically inert form of ATP, being a nitrophenyl derivative conjugated as the phosphate ester of ATP. Photolysis with an appropriate wavelength (347 nm) cleaves in molecule and yields ATP and 2-nitrosoacetophenone [464]. The inert photolable precursor of ATP is incorporated into muscle fibbers. Then ATP is rapidly released by laser pulse photolysis [465]. A simple method of preparing caged (inactive) protein complexes using the amino group and photo-deprotection group of [(nitroveratryl)oxy]chlorcarbamate is described in [515]. Polymerization activity of G-actin is lost when essential lysine residues of G-actin are conjugated with a label-reagent. When caged G-actin is excited by near-UV light, an efficient photo-deprotection reaction run via photoisomerization of the (nitrophenyl)ethyl group of the label. A standard irradiation condition was defined, which led to photoactivation of F-actin by caged G-actin.

18.2 Oxygen Uptake in Mitochondria under Photoexcitation

We studied the effect of illumination with visible and near UV radiation "under the beam" on the mitochondrial respiratory chain [60, 466]. A special setup has been designed for photochemical experiments. The setup consists of a light source (mercury or xenon lamp), a thermal filter (a solution of cupric sulphate), filters, monochromators, a photo-multiplier, amplifiers, a recorder, and a hermetically sealed and temperature controlled quartz mirror cuvette. The cuvette is equipped with a mechanical stirrer, a miniature pH electrode and a membrane Clark electrode for measuring oxygen concentration. The setup allows a simultaneous measurement of fluorescence kinetics, pH and oxygen content during irradiation (Fig. 18.1).

It was found that when a mitochondrial suspension was exposed to visible and near UV light, consumption of oxygen occurred (Fig. 18.2). This consumption stopped just after switching off the light. The illumination intensity was about 5×10^{-5} einstein/1×s. A similar oxygen uptake under visible light in cyanide-treated mitochondria was detected in [467].

In mitochondria oxygen consumption (without adding substrates) induced by the illumination. The addition of ketoglutarate accelerated the process, whereas oxybutyrate, glutamate, and succinate produced no effect. Since rotenone and antimycin only slightly decreased the rate of oxygen photo-uptake, it may be assumed that the photoconsumption of oxygen does not result from functioning of the entire respiratory chain. At the same time, sodium azide decelerated the process about two-fold.

An addition of flavin mononucleotide (FMN) to mitochondria sharply increased photo-uptake of oxygen (Fig. 18.2); in this case the light at 365 nm and 436 nm was much more active than that at 315, 546 and 577 nm. Fig. 18.2 shows

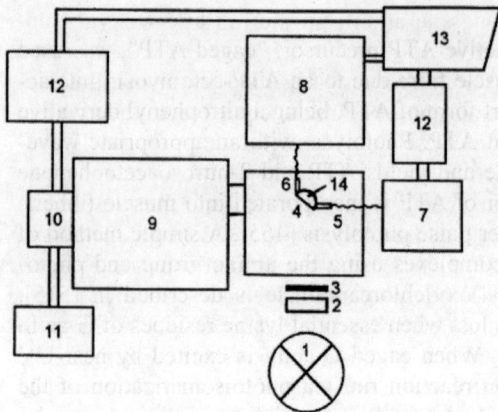

Fig. 18.1. Setup for photochemical experiments. 1, mercury or xenon lamp; 2, thermal filter; 3, filter; 4, cuvette; 5, Clark oxygen electrode; 6, pH electrode; 7, polarograph; 8, pH-meter; 9, monochromator; 10, photo-multiplier; 11, high voltage supply; 12, amplifier; 13, recorder; 14, stirrer

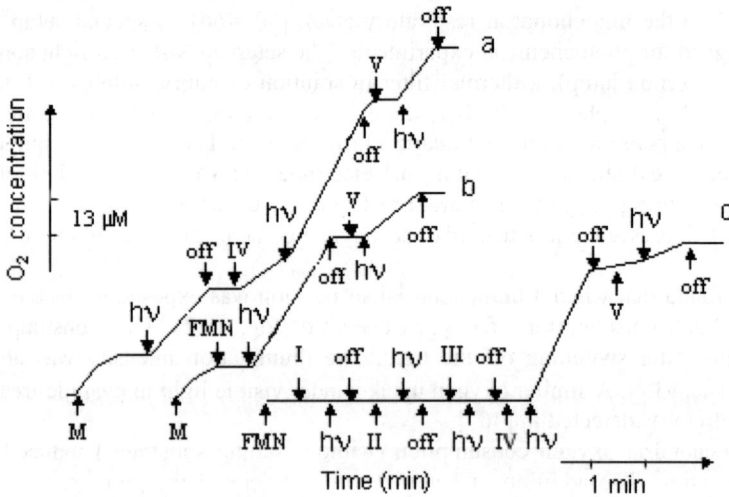

Fig. 18.2. Oxygen uptake induced by: (a) irradiation of mitochondria with visible light and near-UV; (b) irradiation of mitochondria at 436 nm in the presence of 50 μM FMN; and (c) irradiation of the FMN solution (without mitochondria but with adding substrates) at 436 nm [466]. $h\nu$ is switching on the light, OFF is switching off the light. I is addition of succinate, II is addition of glutamate, III is addition of oxybutyrate, IV is addition of ketoglutarate, V is addition of sodium azide. Concentrations: substrates are 4 mM each, mitochondria are 4 mg of protein per ml

oxygen uptake in blue light. Rotenone and antimycin did not affect oxygen uptake, whereas sodium azide suppressed it.

To elucidate the mechanism of oxygen photo-uptake, model systems, i. e., solutions of flavins, NADH, ketoglutarate, phospholipids, etc were studied. When FMN in solution is illuminated in the presence of ketoglutarate, oxygen is consumed (without mitochondria) (Fig. 18.2).

Oxygen photoconsumption without membranes occurs only in the presence of ketoglutarate or NADH. It is strongly inhibited by azide. Succinate, glutamate, and oxybutyrate do not react at all (Fig. 18.2). Oxygen is also consumed during illumination of FMN in the presence of phospholipid suspension (not shown), but such oxygen photo-uptake was not inhibited by azide.

It is known that illumination of flavins results in intersystem crossing to the triplet state [468]. Under aerobic conditions the triplet states are rapidly quenched by oxygen, that results in the production of singlet oxygen (O_2^*) of a high reactivity [6]. It is known that O_2^* easily oxidizes unsaturated fatty acids and phospholipids. Therefore, one of the reactions that should run during illumination of mitochondria is the following:

$$F \xrightarrow{h\nu} F^* \rightarrow F^T \xrightarrow{O_2} F + O_2^* \xrightarrow{L} LO_2$$

where F is the flavoprotein or FMN; F^* and F^T are the excited singlet and triplet states of flavin, respectively; L is the lipids or other oxidizible substrates, and LO_2 are the products of oxidation.

On the other hand, triplet flavins able to efficiently oxidize ketoglutarate or NADH in the absence of oxygen [193–195]. Therefore, in mitochondria the following process should occur:

$$F \xrightarrow{h\nu} F^* \rightarrow F^T \xrightarrow{S} F-H_2 \xrightarrow{O_2} F + H_2O_2$$

where S is the substrate as ketoglutarate or NADH, which is able to donate two electrons to the triplet flavin, and $F-H_2$ is the reduced flavin.

Stimulation of oxygen uptake during illumination of mitochondria of plant seedlings was reviewed in [44]. This photorespiration in plant seedlings has a functional role.

18.3 Oxidation of NADH by Triplet Flavin and Singlet Oxygen

The mechanism of photooxidation of NADH by flavin in solution is considered below in detail [469].

Fig. 18.3 shows the kinetics of oxidation of NADH in solution upon illumination of FMN. Three parameters were simultaneously measured: pH, oxygen content, and flavin fluorescence intensity. NADH was monitored spectrophotometrically as well (Fig. 18.4).

Illumination of FMN in the absence of NADH produced no pH change (Fig. 18.3). Simultaneously oxygen is slowly consumed and the intensity of flavin fluorescence very slowly decreased. The observed changes were due oxygen-dependent photodestruction of flavin. Table 18.1 shows irreversible photobleaching of solutions of FAD, FMN, and FMN with sodium azide or NADH added (initial absorbance was 0.23 at 436 nm). It is seen that FMN undergoes rapid photodestruction. The molecule of FAD, unlike the molecule of FMN includes

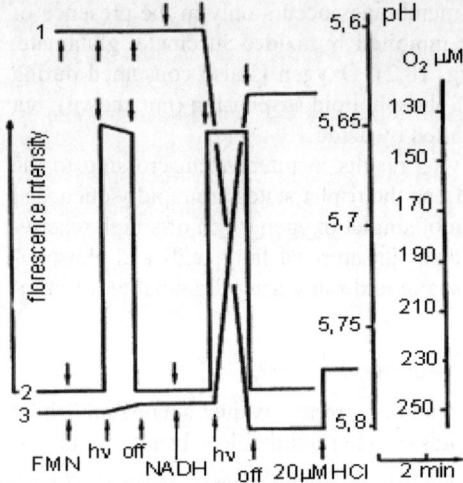

Fig. 18.3. Kinetics of NADH oxidation induced by illumination of FMN [469]: 1) pH change; 2) FMN fluorescence; 3) oxygen uptake. Illumination wavelength is 436 nm. Medium: 2 mM potassium phosphate, pH 5.6; 27 µM FMN, 230 µM NADH. FMN fluorescence is monitored at 520 nm

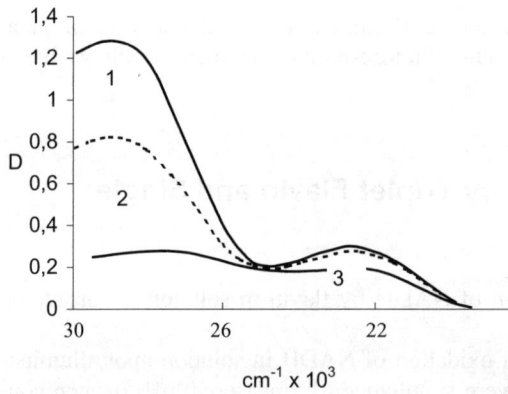

Fig. 18.4. Absorption spectra of a solution of NADH with FMN before and after illumination [469]: 1) before illumination; 2) after illumination at 365 nm for 30 s; 3) after illumination at 436 nm for 30 s. The medium is 20 mM Tris-HCl, pH 7.4

18.3 Oxidation of NADH by Triplet Flavin and Singlet Oxygen

Table 18.1. Photodestruction of flavins in blue light [469]

	FAD	FMN	FMN+NADH	FMN+azide
Initial absorbance	0.23	0.23	0.23	0.23
Final absorbance	0.21	0.15	0.21	0.22
Photodestruction (%)	25	100	25	13

20 mM Tris-HCl, pH 7.4. Concentrations: 20 µM of FMN, 200 µM of oxygen, 2.5 mM of sodium azide, 180 µM of NADH. Illumination wavelength is 436 nm, illumination time is 30 s. Rate of the FMN photodestruction (without any additions) was taken as 100%.

adenine. FAD is almost unbleached, because adenine efficiently quenches the excitation (Chapter 2). An inhibitor of FMN destruction is sodium azide, a well known quencher of singlet oxygen. NADH also prevents the destruction of FMN; NADH being oxidized (Fig. 18.3 and Fig. 18.4). Under anaerobic conditions no photodestruction of FMN is observed.

It can be assumed that NADH in solution is oxidized upon illumination of FMN not only by the reaction

$$FMN^T + NADH \xrightarrow{H^+} FMN - H_2 + NAD^+$$

but also by the singlet oxygen generated by the triplet flavin:

$$FMN^T + O_2 \rightarrow FMN + O_2^* \xrightarrow{NADH + H^+} NAD^+ + H_2O_2$$

where H^+ is put above the arrow, since NADH is a donor of two electrons and one proton, and the second proton is taken from the medium.

Fig. 18.5 shows the pH dependence of oxygen uptake resulting from the photochemical oxidation of NADH by flavin. The reaction is shown to be activated at acidic pH (the fall at pH < 3.5 is due to acidic destruction of NADH and FMN). One of the photo-reaction products is hydrogen peroxide [470].

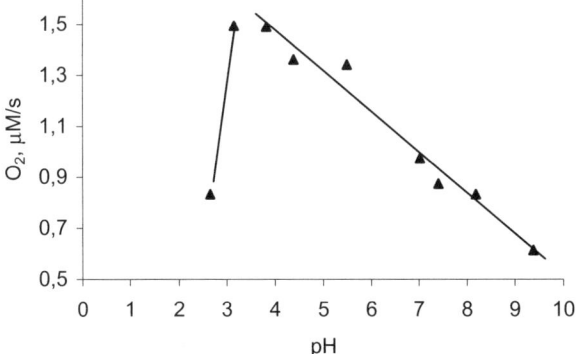

Fig. 18.5. pH dependence of the photooxidation of NADH in the presence of FMN (it was measured by oxygen uptake). Concentrations are: NADH is 180 µM, FMN is 27 µM. Buffer is 20 mM potassium phosphate. Illumination wavelength is 365 nm

Table 18.2. The rates of aerobic and anaerobic photooxidation of NADH in the presence of FMN [469]

	Rate (relative units) under illumination at:	
	436 nm	365 nm
Aerobic conditions	14.3	8.1
Anaerobic conditions	1.3	1.7

2 mM potassium phosphate, pH 5.3, 280 µM NADH, 80 µM FMN.

The reaction rate strongly depends on the presence of oxygen. Table 18.2 shows the rate of NADH oxidation by flavin under aerobic and anaerobic conditions at two different illumination wavelengths. Here the concentration of FMN was increased up to 80 µM so that the reaction rate (measured by alkalinization of the medium) was not limited by flavin concentration. For photoexcitation at 436 nm, the reaction rate under aerobic conditions (200 µM O_2) is much higher than that under anaerobic conditions. With illumination at 365 nm the difference in rates is 5 time less. The 365-nm light is partially absorbed by NADH, which also can be "photo-active", i.e. able to donate electrons more easily upon illumination [471, 472].

Table 18.3 demonstrates the rates of NADH oxidation with and without flavins, in the presence of sodium azide or $NiCl_2$, at different wavelengths and times of illumination. FAD is considerably less photoactive compound than FMN. Azide and $NiCl_2$ sharply decrease NADH oxidation. Illumination at 436 nm is much more efficient than at 365 nm. Illumination of NADH at 365 nm in the absence of flavins results in no photo-oxidation, i.e. the photoexcited NADH has no tendency to be oxidized directly by oxygen (intense UV illumination in the range from 250 nm to 365 nm leads to a slow photo-oxidation of NADH).

Table 18.3. The rate of photochemical oxidation of NADH under aerobic conditions [469]

System	λ, [nm]	Δt, [s]	$\Delta NADH/\Delta t$, [µM/s]
NADH+FMN	365	15	2.26
NADH+FAD	365	30	0.43
NADH+FAD	436	15	1.20
NADH+FMN	436	10	7.20
NADH+FMN+NaN_3	436	30	0.80
NADH+FMN+NaN_3	365	30	0.40
NADH+FMN+$NiCl_2$	365	15	0.33
NADH+FMN+$NiCl_2$	436	15	1.66
NADH	365	30	0.06
NADH	UV	30	0.80
NADH+NaN_3	UV	30	0.67

20 mM Tris-HCl, pH 7.4. Concentrations are: 180 µM of NADH, 20 µM of FAD, 21 µM of FMN, 2.5 mM of sodium azide, 2 mM of nickel chloride; UV-light is within 250 nm to 365 nm; Δt is illumination time.

18.3 Oxidation of NADH by Triplet Flavin and Singlet Oxygen

The data obtained in [469] and discussed above enable to draw the scheme of processes in the solutions of flavins and NADH illuminated with blue light:

$$FMN + h\nu \xrightarrow{k_1} FMN^*$$

$$FMN^* \xrightarrow{k_2} FMN$$

$$FMN^* \xrightarrow{k_3} FMN^T$$

$$FMN^T \xrightarrow{k_4} FMN$$

$$FMN^T + O_2 \xrightarrow{k_5} FMN + O_2^*$$

$$O_2^* \xrightarrow{k_6} O_2$$

$$O_2^* + NADH + H^+ \xrightarrow{k_7} NAD^+ + H_2O_2$$

$$O_2^* + FMN \xrightarrow{k_8} PO_2$$

$$O_2^* + NaN_3 \xrightarrow{k_9} O_2 + NaN_3$$

$$FMN^T + NADH + H^+ \xrightarrow{k_{10}} FMNH_2 + NAD^+$$

Based on concentration dependences of the photo-oxidation rate under various conditions and literature data on the rate of quenching of triplets by oxygen and on the rate of quenching of singlet oxygen by azide, the constants for the main processes were estimated [469]. Table 18.4 summarizes the rate constants and the quantum yields of photo-oxidation.

The quantum yields of photo-oxidation were determined by the same method as in the studies of photo-oxidation of porphyrin [473, 474]; the quantum yield of photo-oxidation for NAPH is expressed as follows:

$$\varphi_{NADH} = \frac{k_7 [NADH] k_5 [O_2] \gamma^T 100\%}{(k_4 + k_5 [O_2])(k_6 + [NADH] k_7 + [FMN] k_8)} \qquad (18.1)$$

where γ^T is the quantum yield of intersystem crossing of flavin to a triplet, NADH=1.8×10^{-4} M, FMN=2×10^{-5} M, $O_2 \approx 2 \times 10^{-4}$ M. The calculated quantum yield of photo-oxidation of NADH by singlet oxygen is about 70%.

Our data on photo-oxidation of NADH in solution by triplet flavin and singlet oxygen were confirmed by Krasnovsky et al. [475], although the yield of photo-oxidation of NADH by singlet oxygen was estimated not more than 30%. Evidence for NADH oxidation by singlet oxygen was also obtained with other photo-sensitizer, i.e., 2-acetonaphthone, by Peters and Rodgers [476]. Our method of 50% azide inhibition of the reactions with participation of triplet and singlet oxygen for the estimation of the rate constants [469] was successfully applied to various systems by other authors [475, 477, 478]. Fritz and Ninnemann [204] have studied the role of singlet oxygen and triplet flavin in photoreactions of nitrate reductase from *Neurospora crassa*.

Table 18.4. Rate constants and quantum yields of photooxidation for the reaction of NADH with flavins and oxygen

Constant	Constant value	Photo-oxidation yield, [%]
k_4	$\ll 10^6 \, s^{-1}$	–
k_5	$\gg 10^9 \, M^{-1} s^{-1}$	–
k_6	$5 \times 10^5 \, s^{-1}$	–
k_7	$2 \times 10^9 \, M^{-1} s^{-1}$	70
k_8	$10^4 \, M^{-1} s^{-1}$	10
k_9	$2 \times 10^9 \, M^{-1} s^{-1}$	0
k_{10}	$1.5 \times 10^7 \, M^{-1} s^{-1}$	20

Photosensitizer is FMN, aerobic conditions, illumination at 436 nm.

Photo-consumption of oxygen in flavoprotein-contained membranes results from the following processes:

$$Fla \xrightarrow{h\nu} Fla^* \to Fla^T \xrightarrow{O_2} Fla + O_2^* \xrightarrow{Lip} LipO_2$$

$$Fla \xrightarrow{h\nu} Fla^* \to Fla^T \xrightarrow{Sub} FlaH_2 \xrightarrow{O_2} Fla + H_2O_2$$

where Fla^* and Fla^T are the singlet and triplet excited flavines, respectively, O_2^* is the singlet oxygen, Sub is the substrate (NADH or ketoglutarate), Lip is the lipid.

The oxidation of NADH by singlet oxygen is important in "dark" systems as well, since in animal cells O_2^* could be formed at destruction of peroxide compounds.

18.4 Conclusion

Photo-uptake of oxygen in suspension of rat liver mitochondria was found. By using inhibitory analysis and studying model systems including NADH, flavins, ketoglutarate, and phospholipids in solution, it has been shown that oxygen photo-uptake is the result of oxidation of these compounds by triplet flavin and singlet oxygen.

19 Light-Dependent Phosphorylation in Mitochondria

Absorption of light by mitochondrial flavoproteins is accompanied by a number of processes: internal conversion, vibrational relaxation etc. Since the quantum yields of fluorescence and phosphorescence of flavin chromophores in flavoproteins are low [70], the main portion of photoexcitation quanta is transformed to vibronic quanta, i. e to the energy of vibrational and orientational relaxation of protein structure, or, in other words, to a local "heating" of the membrane. This can influence on the rate of oxygen uptake and ATP synthesis.

Straub [479] suggested an original hypothesis, according to which the functioning of the respiratory chain is accompanied by the release of phonons (collective vibrational excitations similar to excitons) during the transfer of electrons along the mitochondrial chain. These collective vibrational excitations can reach ATPase and their energy can be used for ATP synthesis. Unfortunately, the phonon hypothesis in its original version was not supported by experimental data, but it contained a rational kernel: a physical mechanism of coupling of electron transfer and ATP synthesis via electron-phonon interactions. Straub assumed that photo-excitation (by illumination) of the components of the respiratory chain in mitochondria might result in ATP synthesis [479].

Most authors studied the effect of illumination on non-photosynthesizing membranes using the following common approach. Membranes or cells were intensively illuminated either in the presence or in the absence of a specially introduced photosensitizer, and then the activities of enzymes were estimated in the dark. Aggarwal et al. [480] studied the effect of light on rat liver mitochondria in the absence of any added sensitizers. It was found that illumination for several hours affects catalytic activities of ATPase, succinate dehydrogenase, NADH dehydrogenase, and activates oxygen uptake, which vanished with time. The experiment consisted of prolonged illumination of mitochondrial suspension, sampling of the illuminated suspension in the dark; subsequent measurements of ATP content, oxygen uptake, enzyme activity, etc, in the dark.

19.1 ATP Synthesis during Illumination

We studied the effect of light on respiration of mitochondria from rat liver in suspension during illumination and found rapid oxygen uptake [466]. Also, it was found that ATP could be synthesized in the light [481–483]. Below we represent

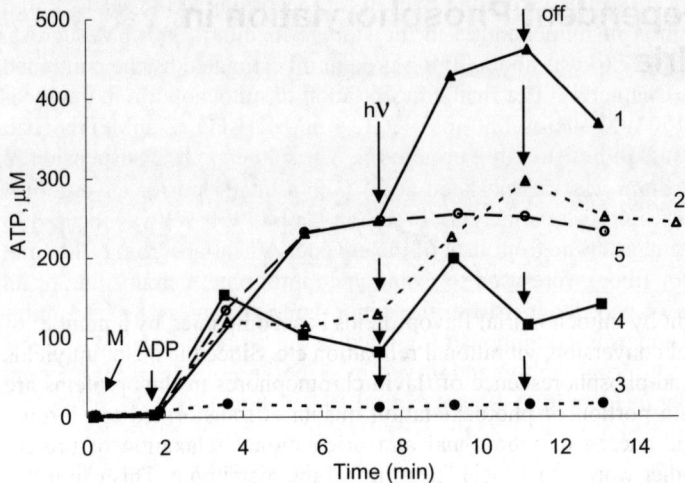

Fig. 19.1. Kinetics of ATP concentration in a suspension of liver mitochondria after the addition of high concentrations of ADP in the dark and in the light. 1) 1000 μM of ADP is added, no substrates; 2) 700 μM of ADP is added, no substrates; 3) 200 μM of ADP is added, no substrates; 4) 200 μM of ADP is added in the presence of 3 mM ketoglutarate; 5) 1000 μM of ADP is added, no substrates, no illumination [482]

our results of the experiments on the light-induced ATP synthesis in mitochondrial suspension.

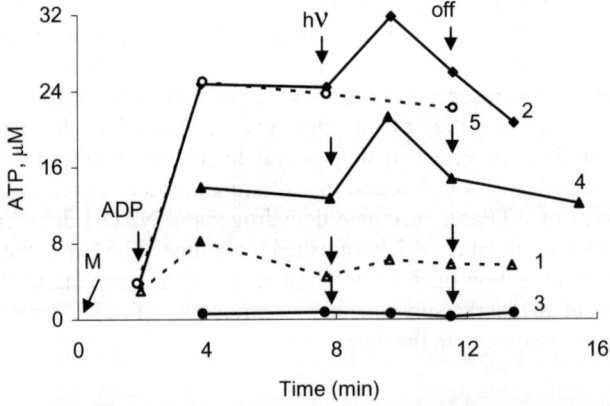

Fig. 19.2. Kinetics of ATP concentration in a suspension of liver mitochondria after the addition of low concentrations of ADP in the dark and in the light: 1) without substrates; 2) in the presence of 3 mM ketoglutarate; 3) in the presence of 3 mM ketoglutarate and 100 μM DCCD; 4) in the presence of 3 mM ketoglutarate and 100 μM dinitrophenol; 5) in the presence of 3 mM ketoglutarate, no illumination. In all cases 50 μM of ADP was added [482]

19.1 ATP Synthesis during Illumination

Mitochondria were isolated from the liver of rats according to [484] and stored on ice. Concentration of mitochondria in the storage medium (300 mM sucrose, 20 mM Tris-HCl, pH 7.6) was about 70 mg protein/ml estimated by the combined UV method [44] (Chapter 4). The final concentration of mitochondria in the incubation medium (150 mM sucrose, 50 mM KCl, 4 mM KH_2PO_4, 5 mM Tris-HCl, pH 7.4) was adjusted to be 1 to 5 mg protein/ml. The mitochondrial suspension in the incubation medium was thermostated at 20°C in a quartz mirror cuvette of 4 ml volume. Visible light from a mercury or xenon lamp (>150 W) was focused to form a wide strip image at the front face of the cuvette. A heat-absorbing filter and a set of glass color filters were used to isolate an appropriate mercury line or an appropriate range of wavelengths from the xenon lamp (Chapter 18). The mitochondrial suspension was continuously agitated using a mechanical stirrer immersed in the cuvette from the top and driven by a motor at a rate of 90 rev/min. The intensity of the incident light did not exceed 10^{-5} einstein l^{-1} s^{-1}.

Addition of mitochondria and ADP to the incubation medium, switching the light on and off, and removing 0.05 ml samples were carried out at 2 min intervals. Each sample was quickly fixed with 4 ml of ice-cold trichloracetic acid. ATP content was quantitated by bioluminescence from the luciferin-luciferase reaction using either Firefly Lantern Extract (Sigma) or a similar extract obtained according to [8]. Respiration of mitochondria was monitored with a Clark membrane electrode; the "respiratory control" was about 3 - 4 when succinate was used as the substrate.

When high concentrations of ADP were added to the mitochondrial suspension in the absence of any substrates of the electron transport chain, phosphorilation occurred (Fig 19.1). This phosphorylation is believed to be determined by two factors: oxidation of endogenous substrates and the activity of adenylate kinase [448]. Adding ADP led to rapid phosphorylation, after which ATP content reached a constant level or decreased due to hydrolysis. Illumination resulted in a drastic increase in the ATP content (Fig. 19.1). The result of this light-dependent phosphorylation strongly depends on the concentration of added ADP. After 2–4 min of illumination the level of ATP stopped increasing or even started to fall. Switching off the light resulted in cessation of light-dependent phosphorylation. In the dark the level of ATP could either decrease or keep constant.

When the concentration of added ADP was lowered from 1000 or 700 μM to 200 μM, dark and light-dependent phosphorylations were drastically decreased [481, 482]. Addition of succinate or another respiratory chain substrate to the incubation medium prior to the addition of ADP accelerated dark phosphorylation, but it had no effect on the rate of light-dependent phosphorylation. The use of ketoglutarate instead of succinate and addition of only 200 μM of ADP resulted in a sharp increase in dark and light-dependent phosphorylations (Fig. 19.1).

Phosphorylation occurs also at those concentrations of ADP, at which the activity of adenylate kinase cannot be high (Fig. 19.2). In this case, dark phosphorylation and light-dependent phosphorylation markedly increase in the presence of ketoglutarate. During the first two minutes of illumination the concentration of ATP sharply rises, and after that it remains constant or even falls. After switching off the light the level of ATP either remains constant or decreases further.

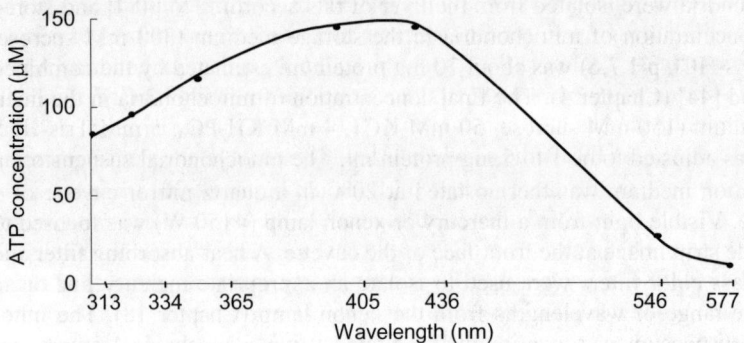

Fig. 19.3. Spectrum of action of light-dependent phosphorylation. The concentration of added ADP is 1 000 μM. Illumination time is 2 min [482]

The use of ATP-synthetase inhibitors as dicyclohexylcarbodiimide (DCCD) or oligomycin prior to the addition of ADP results in a complete depression of both dark phosphorylation and light-dependent phosphorylation. Fig. 19.2 shows the data obtained with DCCD. Inhibitors of the respiratory chain (rotenone, antimycin, sodium azide) suppressed dark and light-dependent phosphorylation. The uncoupler 2,4-dinitrophenol significantly decreased dark phosphorylation, but it did not influence the rate of light-dependent phosphorylation (Fig. 19.2).

Since the strongest light-dependent phosphorylation was observed with 1000 μM ADP, the action spectrum, i.e., the dependence of light-synthesized ATP content on the excitation wavelength, was measured with the addition of 1000 μM ADP. A mercury lamp was used as the light source. The lamp had intense radiation lines at 313, 334, 365, 405, 436, 546, and 577 nm (other lines were not used because of difficulties related to the monochromatization and because of the low intensities). The points in Fig. 19.3 were calculated from the experimental values obtained after 2 min of illumination. The data were corrected for the known intensities of the mercury lines and for the transmission of the filters. It can be seen that the peak in the action spectrum is near 436 nm, and the shape of the spectrum is similar to that of the absorption spectrum of oxidized flavin [60], although the action spectrum is significantly shifted to short wavelengths as compared to the absorption of flavin. This may be related to the fact that not only flavoproteins but also cytochromes [448], which have an intense band at 400–430 nm, are responsible for phosphorylation.

During illumination of water-soluble non-purified mitochondrial ATPase, ATP synthesis was observed [485], which is in agreement with our data for mitochondria [481, 482]. Irradiation of rat liver mitochondria using a He-Ne laser generated an extra electrochemical potential as well as an increase in the synthesis of ATP [486]. The ATP synthesis in mitochondria can be inhibited or promoted by UV irradiation, producing active forms of oxygen [487]. It was shown a light stimulation of the ATP synthesis in mitochondria in wheat leaf protoplasts [488] and in plant seedling [44]; such a synthesis has a biological function.

The evidence for light-dependent ATP synthesis in liver mitochondrial suspensions is virtually analogous to the findings obtained with chromophores of purple bacteria, chloroplasts, "chimeras" of bacteriorhodopsin and mitochondrial ATPase, etc [37, 43, 60, 460, 461, 489, 490]. Increasing the amount of bacteriorhodopsin in the proteoliposomes at a constant the ATPase concentration led to a large increase in the rate of ATP synthesis whereas the magnitude of the transmembrane ΔpH remained the same or decreased [490]. It should be noted that there is a parallel between our findings and the data obtained in the experiments with light-dependent sorption-desorption of cGMP in external segments of retinal rods without the participation of enzymes. If GTP is added to the external segments of rods, it binds cGMP. When the latter are illuminated, cGMP is desorbed [34]. The light-dependent sorption-desorption of substrates seems to be a common property of very different membrane systems.

19.2 Thermal Coupling between ATP Synthesis and Electron Transfer

After a part of the added ADP is phosphorylated in the dark, the system reaches a stationary state, in which the synthesis and hydrolysis of ATP proceed at relatively close rates. Illumination shifts the ratio ATP/ADP toward the production of ATP. What is the mechanism of such phosphorylation?

After photoexcitation of flavins and ferrous porphyrins a rapid vibrational relaxation and internal conversion take place [60, 491]. Using the method of molecular dynamics Henry et al. [491] studied the rate of energy withdrawal from the heme in cytochrome c after the introduction of excess energy of ~200 or ~300 kJ/mol, which is equivalent to the absorption of a photon with a wavelength of 530 or 350 nm, respectively. According to their data, about one half of the energy is transferred to the environment during several picoseconds, whereas the remainder of the energy is transferred within 20–40 ps. Photoexcitation results in an instant 500–700 K increase in the heme temperature upon the conversion of one photon to vibrational excitations [491]. In [492] experiment proved that photoexcitation of heme proteins produced a successive "heating" of the heme, the protein globule, and the medium (on the time scale from tens of ps to several ns). Along with vibrational relaxation and internal conversion, intersystem crossing to a triplet is also typical of flavins. However, a triplet flavin in flavoproteins is usually converted to the ground state without triplet emission [60]. Thus, most of the photoexcitation quanta in cytochromes and flavoproteins are transformed to vibrational excitations, i.e. they produce local heating of the proteins and their surroundings.

It may be assumed that the vibrational excitation quanta formed after photoexcitation of mitochondrial flavoproteins and cytochromes are trapped by ATP-synthetase with some efficiency, and this induces desorption of ATP from the active center. This is the reason why the reaction shifts toward ATP formation. The wave numbers of the vibrational quanta are ranged inside the IR region between

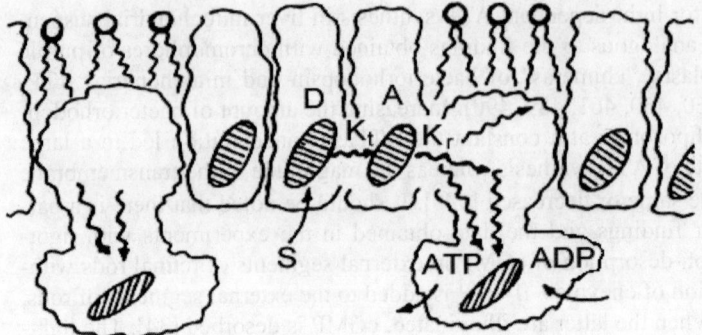

Fig. 19.4. Thermal model of coupling in membranes [493]. K_t is the rate constant of electron transfer, K_r is the relaxation constant, D and A are donors and acceptors of electrons and energy

~4000 cm^{-1} and ~1000 cm^{-1} that correspond to the wavelengths of 2500–10000 nm. Therefore, one vibrational wave in the mitochondrial respiratory chain overlaps all ATPase molecules.

This model correlates well with the results of [326, 494], which suggest that no energy is required for the phosphorylation in the active center of ATP-synthetase, but the desorption of ATP requires energy. Earlier [495, 496] it was shown that ATP in preparations of water-soluble ATPases from mitochondria and chloroplasts formed in the absence of any source of energy. The formed ATP remains bound to the enzyme and is inaccessible to hexokinase.

It is known that long-term illumination damages mitochondria. The photochemical reactions with the participation of triplet flavin and singlet oxygen destroy mitochondrial membranes [60]. After prolonged illumination of mitochon-

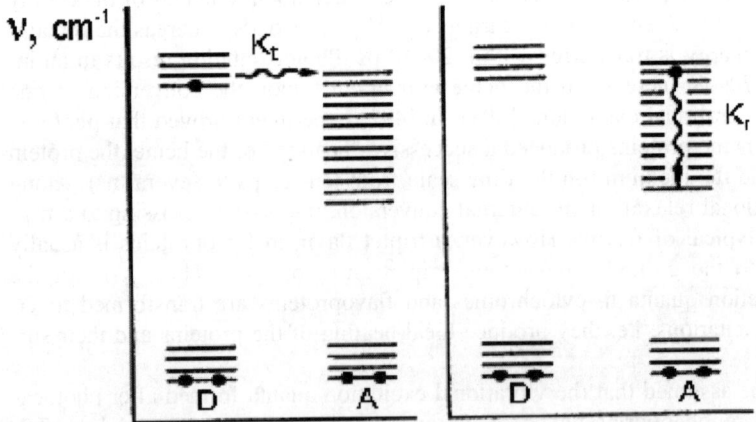

Fig. 19.5. Diagram of electron transfer and vibrational relaxation in the components of the respiratory chain [493]. Notation is the same as in Fig. 19.4

dria, ATPase hydrolytic activity is drastically increased [480]. ATP hydrolysis in our experiments was activated by illumination of mitochondria for longer than 2–3 min.

The light may influence the activity of ATP-synthetase in three mitochondrial enzyme systems: adenylate kinase, ketoglutarate dehydrogenase and respiratory ATPase.

Fig. 19.4 shows the scheme of functioning of electron transport chain based on thermal coupling "pumped" with light quanta or substrate energy, S.

Light-dependent phosphorylation in mitochondria seems to have much in common with dark phosphorylation and with photophosphorylation in chloroplasts. During the electron transfer between isoenergetic levels of a pair of components of the electron transport chain, vibrational relaxation inevitably takes place. Relaxations during dark electron transfer are virtually the same as the vibrational relaxation or internal conversion after photoexcitation. Fig. 19.5 shows the scheme of relaxation in a pair of components of the chain. This problem is considered in more detail in the review [493]. The "thermal synthesis" of ATP in membranes has attracted considerable interest [497, 498].

19.3 Conclusion

Illumination of mitochondria can result in light-dependent phosphorylation, sensitive to inhibitors. It is suggested that light produce a shift in the enzymatic reaction toward ATP synthesis. This effect can be result from the desorption of ATP at the moment of local heating of the membrane due to the internal conversion and vibrational relaxation in mitochondrial flavoproteins and cytochromes.

REFERENCES

1. Horrocks WDW (1993) Luminescence spectroscopy. In: Methods in Enzymology, 226, N PC, pp 495–538.
2. Warner IM, McGown LB (1988) Molecular fluorescence, phosphorescence, and chemiluminescence spectrometry. Anal Chem, 60, N 12: 162R–175R.
3. Itzhaki LS, Evans PA, Dobson C.M, Radford S.E (1994) Tertiary interactions in the folding pathway of hen lysozyme. Kinetic studies using fluorescent probes. Biochemistry, 33, N 17: 5212-5220.
4. Veeraraghavan S, Holzman TF, Nall BT (1996) Autocatalyzed protein folding. Biochemistry, 35: 10601–10607.
5. Rothwart DM, Scheraga HA (1996) Role of non-native aromatic and hydrophobic interactions in the folding of hen egg white lysozyme. Biochemistry, 35: 13797–13807.
6. Terenin AN (1967) Photonics of Dye Molecules and Related Organic Compounds (in Russian). Leningrad: Nauka.
7. Sarzhevsky AM, Sevchenko AN (1971) Anisotropy of Light Absorption and Emission by Molecules (in Russian). Minsk: Belorussian State University.
8. Rabek JF (1985) Experimental Methods in Photochemistry and Photophysics (in Russian). vol 1 and 2. Moscow: Mir.
9. Riehl JP, Richardson FS (1993) Circularly-polarized luminescence. In: Methods in Enzymology, 226: 539–553.
10. Levshin LV., Saletskiy AM (1989) Luminescence and Its Measurements (in Russian). Moscow: MGU Publ.
11. Schulman SG (1985) Luminescence Spectroscopy: in overview. In: Molecular Luminescence Spectroscopy. Methods and Applications: Part 1 (Chemical Analysis, vol 77, ed. by P.J.Elving and J.P.Winefordner) N-Y : John Wiley & Sons.
12. Levshin LV, Saletskiy AM (1994) Optical Methods in Studies of Molecular Systems (in Russian), part 1. Moscow: MGU Publ.
13. Baeyens WRG et al (1991) (eds) Luminescence Techniques in Chemical an Biochemical Analysis. N–Y.: Marcel Dekker.
14. Dewey TG (1991) (eds) Biophysical and Biochemical Aspects of Fluorescence Spectroscopy. N-Y : Plenum.
15. Konev SV (1965) Electron-Excited States of Biopolymers (in Russian). Minsk: Nauka i tekhnika.
16. Barenboim GM, Domansky AN, Turoverov KK (1966) Luminescence of Biopolymers and Cells (in Russian). Moscow: Nauka.
17. Burstein EA (1977) Intrinsic protein luminescence. In: Biofizika, vol 7, Itogi nauki i tekhniki (in Russian). Moscow: VINITI.
18. Chernitsky EA (1972) Luminescent and Structural Lability of Proteins in a Solution and Cell (in Russian). Minsk: Nauka i tekhnika.

19. Vladimirov YuA, Dobretsov GE (1980) Fluorescent Probes in Studies of Biological Membranes (in Russian). Moscow: Nauka.
20. Subcellular Biochemistry (1988) vol 13. Fluorescence Studies on Biological Membranes (eds by Hilderson). New York: Plenum Press.
21. Lin T-L, Dowben RM (1983) Nanosecond pulse fluorimetry of proteins. In: Excited States of Biopolymers (eds by R.P.Steiner). New York: Plenum Press, pp 59–115.
22. Demchenko AP (1988) Luminescence and Dynamics of Protein Structure (in Russian), Kiev: Naukova dumka.
23. Zolin VF, Koreneva LG (1980) Rare-Earth Probe in Chemistry and Biology (in Russian). Moscow: Nauka.
24. Steiner RF, Kubota Y (1983) Fluorescent dye-nucleic acid complexes. In: Excited States of Biopolymers (ed. by R.F.Stemer). New York: Plenum Press, pp 203–254.
25. Bayley PM, Dale RE (1985) (eds) Spectroscopy and the Dynamics of Molecular Biological Systems. London: Academic Press.
26. Horrocks WDW (1993) Luminescence spectroscopy. In: Methods in Enzymology, 226, pp 495–538.
27. Ladokhin AS, Selsted M, White SH (1997) Bilayer interactions of indolicidin, a small antimicrobial peptide rich in tryptophan, proline, and basic amino acids. Biophys J, 72: 794–805.
28. Warner IM, Soper SA, McGown LB (1996) Molecular fluorescence, phosphorescence, and chemiluminescence spectrometry. Anal Chem, 68: 73R–91R.
29. Labas YuA (1973) On the mechanisms of calcium-activated bioluminescence of *Ctenophoras*. In: Biophysics of a Living Cell (in Russian). N 4. Pushchino: Acad Sci of the USSR, pp 83–109.
30. Baranova NA, Danilov VS, Egorov NS (1980) Peculiarities of functioning of bioluminescence chain. Dokl Acad Nauk (in Russian), 252, N 4: 1009–1012.
31. Ward WW, Cormier MJ (1978) Energy transfer via protein-protein interaction in Renilla bioluminescence. Photochem Photobiol, 27, N 4: 389–396.
32. Hastings JM, Balny C, Douzou P (1976) The oxygenated luciferase- flavin intermediate. In: Flavins and Flavoproteins (eds by T.P.Singer). Amsterdam Elsevier, pp 53–61.
33. Henry JP, Michelson AM (1978) Bioluminescence: physiological control and regulation at the molecular level. Photochem Photobiol, 28: 293–310.
34. Vladimirov YuA, Potapenko AY (1989) Physicochemical Basis of Photobiological Processes (in Russian). Moscow: Vysshaya shkola.
35. Krasnovsky AA (1974) Transformation of Light Energy by Photosynthesis. Bakh Lectures (in Russian). Moscow: Nauka.
36. Gurinovich GP, Sevchenko AN, Solov'ev KN (1968) Spectroscopy of Chlorophyll and Related Compounds (in Russian). Minsk: Nauka i tekhnika.
37. Rubin AB (1987) Biophysics (in Russian), vol 1 and vol 2. Moscow Higher school.
38. Letokhov VS, Matveets YuA, Sharkov AV, Shuvalov VA, Borisov AYa, Proskuryakov II, Nikogosyan DN, Djagarov BM, Chirvony VS, Khoroshilova EV, Zavil'gelsky GB (1987) Laser Picosecond Spectroscopy and Photochemistry of Biomolecules (in Russian). Moscow: Nauka.
39. Borisov AY (1987) Energy migration during photosynthesis. Biofizika (in Russian), 32, N 6: 1046–1061.
40. Kuzmin VA, Chibisov AK (1973) Reaction of primarily-oxidized form of chlorophyll. In: Problems of Biophotochemistry (in Russian). Moscow: Nauka, pp 44–50.

41. Shuvalov VA, Litvin FF (1973) Energetics of luminescence centers of photosynthesizing organisms and their relationship with the reaction centers of photosynthesis: relationship between chemiluminescence and changes in the quantum yield of fluorescence. In: Problems of Biophotochemistry (in Russian). Moscow: Nauka, pp 180–185.
42. Kononenko AA (1973) Photoinduced electron transfer reactions in purpure bacteria. In: Problems of Biophotochemistry (in Russian). Moscow: Nauka, pp 148–154.
43. Hartmann R, Oesterhelt D (1977) Bacteriorhodopsm-mediated photophosphorylation in Halobacterium halobium. Eur J Biochem, 77, N 2: 325–335.
44. Shakhov AA (1993) Photoenergetics of Plants and Yield (in Russian). Moscow: Nauka.
45. Trettnak W (1993) Optical sensors based on fluorescence quenching. In: Fluorescence Spectroscopy: New Methods and Applications (eds by Wolfbeis O.S.). Springer-Verlag, Berlin, pp 79–89.
46. Busch M, Hobel W, Polster J (1993) Software Fiacre: Bioprocess monitoring on the basis of flow-injection analysis using simultaneously a urea optode and a glucose luminescence sensor. J Biotechnol, 31, N 3: 327–343.
47. Wolfbeis OS (1993) Progress in optical chemical sensors and biosensors. In: The 8th International Conference on Solid-State Sensors and Actuators. Yokohama: PACIPICO, pp 542–546.
48. Wolfbeis OS (1993) Optrodes for measuring enzyme activity and inhibition. In: Uses of Immobilized Biological Compounds (eds by G.G.Guibault and M.Mascini). Adv Sci Ser, part E: Appl Sci, vol.252. Dortrecht: Kluwer Acad Publ, pp 335–344.
49. Vekshin NL (1988) Polarized energy transfer in macromolecules. In Proceedings of the USSR. Conference on Spectroscopy of Biopolymers. Kharkov, pp 66–67.
50. Phillips D (1994) Luminescence lifetimes in biological systems. Analyst, 119, N 4: 543–550.
51. Shinitzky M (1972) Effect of fluorescence polarization on fluorescence intensity and decay measurements. J Chem Phys, 56, N 12: 5079–5981.
52. Stevens B (1971) Photoassociation in aromatic systems. In Adv in Photochem, vol 8 (eds by J.N Pitts, IR. Et al), pp 161–226
53. Barltrop JA, Coyle JD (1978) Excited States in Organic Chemistry (in Russian). Moscow: Mir.
54. Birks JB (1970) Excimer fluorescence of aromatic compounds. In: Progress in Reaction Kinetics, vol 5 (eds by G. Porter). Oxford Perg Press, pp 181–272.
55. Birks JB (1970) Photophysics of Aromatic Molecules. London: Wiley.
56. Guillet J (1985) Polymer Photophysics and Photochemistry. Cambridge University Press.
57. Kapinus EI (1988) Photonics of Molecular Complexes (in Russian). Kiev. Naukova dumka.
58. Cantor CR, Shimmel PR (1984) Biophysical Chemistry (in Russian), vol 2. Moscow: Mir.
59. Rubin AB (1988) (eds). Contemporary Methods of Biophysical Studies: Practice on Biophysics (in Russian), MGU, Moscow.
60. Vekshin NL (1988) Photonics of Biological Structures (in Russian). Pushchino: Acad Sci USSR.
61. Neporent BS, Bakhshiev NG (1958) Intensities in spectra of polyatomic molecules. Optika i Spektroskopiya (in Russian), 5: 634–645.

62. Wada A (1972) Dichroic spectra of biopolymers oriented by flow. Appl Spectr Rev, 6: 1–30.
63. Vekshin NL (1999) Screening hypochromism of chromophores in macromolecular biostructures. Biofizika (in Russian), .44, N 1: 45–55.
64. Gillespie RJ, Hargittai I (1991) The VSEPR Model of Molecular Geometry, Allyn and Bacon, Boston.
65. Antipin MY (1990) Low-temperature X-ray analysis: possibilites in resolution of chemical problems, Uspechi Chimii (in Russian), 59: 1052–1084.
66. Dvorkin GA, Krinsky VI (1961) Absorption of light by the solution of deoxyribonucleic acid oriented in electric field. Dokl Akad Nauk USSR (in Russian), 140, N 4: 942–945.
67. Lakowicz J (1983) Principles of Fluorescence Spectroscopy. Plenum Press, N-Y.
68. Vekshin NL (1987) Screening hypochromism in chromophore stacks. Optika i Spektroskopiya (in Russian), 63, N 3: 517–519.
69. Vekshin NL (1989) Screening hypochromism of biological macromolecules and suspensions. J Photochem Photobiol, B: Biol, 3, N 4: 625–630.
70. Udenfriend S (1965) Fluorescent Analysis in Biology and Medicine (in Russian). Moscow: Mir.
71. Castanho MARB. Santos NC, Loura LMS (1997) Separating the turbidity spectra of vesicles from the absorption spectra of membrane probes and other chromophores. Eur Biophys J, 26: 253–259.
72. Weissbluth M (1971) Hypochromism. Quart Rev Biophys, 4, N 1: 1–34
73. Lokey RS, Iverson BL (1995) Synthetic molecules that fold into a pleated secondary structure in solution. Nature, 375: 303–305.
74. Tinoco I (1960) Hypochromism in polynucleotides. J. Am. Chem. Soc., 82, N 18: 4785–4790.
75. Bolton HC, Weiss JJ (1962) Hypochromism in the ultra-violet absorption of nucleic acids. Nature, 195: 666–668.
76. Nesbet RK (1964) Theory of hypochromism. Molec Physics, 7, N 3: 211–221.
77. Rhodes W (1961) Hypochromism and other spectral properties of helical polynucleotides. J Am Chem Soc, 83: 3609–3617.
78. Bullough RK (1968) Complex refractive index and a two-band model in the theory of hypochromism. J Chem Phys, 48: 3712–3722.
79. De Voe H (1964) Optical properties of molecular aggregates; classical model of electronic absorption and refraction. J Chem Phys, 41: 393–400.
80. Fowler GN (1966) On the theory of hypochromism. Molec Physics, 11: 31–36.
81. Liver N, Nitzan A, Amirav A, Jortner J (1988) The effect of small cluster environment on molecular oscillator strengths and spectra. J Chem Phys, 88, N 6: 3516–3523.
82. Bakhshiev NG (1972) Spectroscopy of Inter-molecular Interactions (in Russian). Leningrad: Nauka.
83. Steinberg IZ, Anglister J (1981) Light scattering by chromophores at their absorption bands. Ann N-Y Acad Sci, 366: 125–139.
84. Shaganov II, Libov VS (1975) Observation of dipole-dipole interactions in absorption spectra of different condensed media. In: Spectrochemistry of Intra- and Intermolecular Interactions (in Russian), N 1, LGU, Leningrad, pp 51–61.
85. Solvatochromism: Problems and Methods (1989) (in Russian, eds by N.G.Bakhshiev). Leningrad: LGU.

86. Savostyanova MV (1963) Interaction of dyes with high-molecular substances, Uspekchi Khimii (in Russian), 32: 1233–1269.
87. Seyama F, Akahori K, Sakata Y, Misumi S, Aida M, Nagata G (1988) Synthesis and properties of purinophanes: relationship between the magnitude of hypochromism and stacking geometry of purine rings. J Am Chem Soc, 110: 2192–2201.
88. Doyama K, Higashii T, Seyama F, Sakata Y, Misumi S (1988) Synthesis, structure and hypochromism of pyridinopurinophanes. Bull Chem Soc Jpn, 61: 3619–3627.
89. Ts'o POP (1974) Dinucleoside monophosphates, dinucleotides and oligonucleotides. In: Basic Principles in Nucleic Acid Chemistry. POP Ts'o (eds), vol 2. New York: Acad.Press, pp 305–350.
90. Sevostyanova LV (1963) Interaction of dyes with high-molecular substances. Uspekhi Khimii (in Russian), 32, N 10: 1233–1269.
91. Frank-Kamenetsky MD (1967) Spectrophotometry in the field of electronic transitions. In: Physical Methods for Study of Proteins and Nucleic Acids (in Russian). Moscow: Nauka, pp 113–131.
92. Baranova EG (1962) Study of rhodamine association in water solutions, Optica i Spectroscopiya (in Russian), 13: 683–689.
93. Duysens LNM (1956) The flattening of the absorption spectrum of suspensions, as compared to that of solutions. Biochim Byophys Acta, 19: 1–12.
94. Bell LN (1963) Remarks on the coefficient and the absorption spectrum of nondispersing heterogeneous media. Biofizika (in Russian), 8, N 5: 629–632.
95. Bell LN (1965) The peculiarities of absorption Spectrophotometry of biological objects. Biofizika (in Russian), 10: 374–385.
96. Sherudilo AI (1968) Determination of quantity of light-absorptive substance during cytophotometry of highly heterogeneous objects. Biofizika (in Russian), 13, N 4: 741–744.
97. Papageorgiou G (1971) Absorption of light by non-refractive spherical shells. J Theor Biol, 30: 249–254.
98. Fukshansky L (1978) On the theory of light absorption in non-homogeneous objects. J Math Biol, 6: 177–196.
99. Latimer P (1984) A Wave-optics effect which enhances light absorption by chlorophyll in vivo. Photochem Photobiol, 40, N 2: 193–199.
100. Latimer P (1983) The deconvolution of absorption spectra of green plant materials – improved corrections for the sieve effect. Photochem Photobiol, 38, N 6: 731–734.
101. Fukshansky L (1987) Absorption statistics in turbid media. J Quant Spectrosc Radiat Transfer, 38: 389–406.
102. Richter T, Fukshansky L (1994) Authentic in vivo absorption spectra for chlorophyll in leaves as derived from in situ and in vitro measurements. Photochem Photobiol, 59, N 2: 237–247.
103. Fukshansky L, Remisovsky AM, McClendon J, Ritterbusch A, Richter T, Mohr H (1993) Absorption spectra of leaves corrected scattering and distributional error: a radiative transfer and absorption statistics treatment. Photochem Photobiol, 57, N 3: 538–555.
104. Smirnov LV (1952) Anisotropic absorption of light and orientation of electronic oscillators in dye molecules. Dokl Akad Nauk USSR (in Russian), 82, N 2: 237–240.
105. Ashitkova NS, Murav'ova NL, Chernyakovsky PP (1986) Electrochromism as a method of spectral analysis. In: Molekularnaya Spektroskopiya (in Russian), Issue 7. Leningrad: LGU, pp 214–232.

106. Chernyakovsky PP (1963) Application of electrochromism in studies of bio-and synthetic polymers. In: Biofizika (in Russian), vol 15, Itogi nauki i tekhniki. Moscow: VINITI.
107. Michelson A (1966) Chemistry of Nucleosides and Nucleotides (in Russian). Moscow: Mir.
108. Garcia-Rubio L (1987) The effect of molecular size on the absorption spectra of macromolecules. Macromol., 20: 3070-3075.
109. Volkenstein MV (1975) Molecular Biophysics (in Russian). Moscow: Nauka.
110. Davydov AS (1968) Theory of Molecular Excitons (in Russian). Moscow: Nauka.
111. Buch CA (1974) Ultraviolet spectroscopy, circular dichroism and optical rotatory dispersion. In: Basic Principles in Nucleic Acid Chemistry. P.Ts'o (eds), vol 2. New York: Acad Press, pp 91–169.
112. Gueron M, Eisinger J, Lamola AA (1974) Excited states of nucleic acids. In: Basic Principles in Nucleic Acid Chemistry, vol 1, P.Ts'o (eds), Acad Press, N–Y, pp 352–371.
113. Leng M., Felsenfeld G (1966) A study of polyadenylic acid at neutral pH. J Mol Biol, 15: 455-465.
114. Brahms J, Michelson AM, Van Holde KE (1966) Adenylate oligomers in single- and double-strand conformation. J Mol Biol, 15: 467–488.
115. Morcillo J, Gallego E, Peral F (1987) A critical study of the application of ultraviolet spectroscopy to the self-association of adenine, adenosine and 5'-AMP in aqueous solution. J Mol Struct., 157: 353–369.
116. Listsov VN, Suhorukov BI, Blumenfeld LA, Moshkovskii YS, Petuhov VA (1962) Spectroscopic study of DNA at the 200-nm band. Biofizika (in Russian), 7, N 6: 662–663.
117. Vekshin NL (1998) Energy migration along DNA, detected by sensitized fluorescence of intercalating dye Zh Prikl Spektr (in Russian), vol 65, N 5: 794–798.
118. Borisova OF, Surovaya AN (1973) Application of fluorescent dyes for study of the nucleic acids structure. In: Molecular Biology, vol 1, Itogi nauki i techniki (in Russian), Moscow: VINITI, pp 141–196.
119. Vekshin NL, Vincent M, Gallay J (1993) Tyrosine hypochromism and absence of tyrosine-tryptophan energy transfer in phospholipase A2 and ribonuclease Tl. Chem Phys, 171: 231–236.
120. Urry DW, Pettogrew JW (1967) Model systems for interacting heme moieties. The ferriheme octapeptide of cytochrome C. J Am Chem Soc, 89: 5276–5283.
121. Alpern M, Fulton AB, Baker BN (1987) Self-screening of rhodopsin in rod outer segments. Vision Res, 27, N 9: 1459–1470.
122. Stern ES, Timmons CJ (1966) Gillam and Stern's Introduction to Electronic Absorption Spectroscopy in Organic Chemistry (in Russian). Edward Arnold Publ Ltd, London.
123. Chandross EA, Ferguson J, McRae EG (1966) Absorption and emission spectra of anthracene dimers. J Chem Phys, 45: 3546–3553.
124. Chandross EA, Ferguson J (1966) Photodimerization of crystalline anthracene. J Chem Phys, 45: 3564–3567.
125. Kozel SP, Lashkov GI (1982) Luminescence studies of bi-molecular processes in solutions of anthracene-containing polymers with polymethyl-methacrylate. In: Excited Molecules: Kinetics of Conversions (in Russian) Leningrad. Nauka, pp 188–201.
126. Chromy V, Voznicek I (1982) Total Protein. Praha: Chemapol.

127. Scopes RK (1982) Protein Purification. Springer-Verlag, Berline.
128. Dann M, Maddy AH (1979) Mitochondrial proteins In: Biological Analysis of Membranes (eds by A.H.Maddy, in Russian). Moscow: Mir Publ, pp 176–222.
129. Aliverdieva DA, Sholts KA (1984) Quantitative estimation of general protein in mitochondria with cumassi. Prikladnaya Biochimiya i Microbiologiya (in Russian), 20: 823–830.
130. Shishkin SS (1982) Using the dyes for protein estimation in solutions. Voprosy Medic Chim (in Russian), 28: 134–139.
131. Thorne CJR (1978) Protein estimation In: Techniques in Protein and Enzyme Biochemistry. vol 104, pp 1–18, Elsevier–North Holland Biomed. Press.
132. Kochetov GA (1980) Practical Guide on Enzymology (in Russian). Moscow: Vysshaya shkola.
133. Meshkova NP, Severin SE (1979) Laboratory Course of Biochemistry (in Russian). Moscow. MGU Publ.
134. Vekshin NL (1988) Spectrophotometric estimation of protein content in standard biological suspensions. Biologicheskiye Nauki (in Russian), N4: 107–111.
135. Ritov VB, Kozlov YuP, Murzakhmetova MK (1983) Molecular Organization and Functional Properties of Transport Ca^{2+}-dependent Adenosinetriphosphatase of Sarcoplasmic Reticulum (in Russian). Irkutsk: Irkutsk State University.
136. Kushmanova OD, Ivchenko GM (1983) Guide to Laboratory Manual of Biological Chemistry (in Russian). Moscow: Medicine Publ.
137. Dobretsov GE, Vekshin NL, Vladimirov YuA (1978) Difference in spacial organization between the protein-lipid complexes of endoplasmic reticulum membranes from liver and sarcoplasmic reticulum. Dokl Akad Nauk USSR (in Russian), 239, N 5: 1241–1244.
138. Parker S (1972) Photoluminescence of Solutions (in Russian). Moscow: Mir.
139. Soengas MS, Mateo CR, Salas M, Acuna AU, Gutierrez C (1997) Structural features of single-stranded DNA-binding protein. J Biol Chem, 272, N 1: 295–302.
140. Vekshin NL (1988) Microscreening and microreabsorption of luminescence in heterogeneous systems. Zh Prikl Spektros (in Russian), 49, N 1: 31–35.
141. Oelkrug D, Mammel U, Brun M, Gunther R, Uhl S (1993) Fluorescence spectroscopy on light scattering materials. In: Fluorescence Spectroscopy: New Methods and Applications, Wolfbeis O.S. (eds), Springer-Verlag, Berlin, pp 65–78.
142. Kurdoglyan MS (1993) Quantum theory of resonance intermolecular energy transfer in micron-sized dielectric particles. Opt i Spektr (in Russian), 75, N 4: 826–829.
143. Vekshin NL (1986) On the increase in acceptor luminescence upon reabsorption of donor luminescence. Zh Prikl Spektr (in Russian), 44, N6: 961–965.
144. Zhevandrov ND, Gorshkov VK, Yashin VA (1971) Study of energy transfer in solutions of some aromatic hydrocarbons and lumogens. Zh Prikl Spektr (in Russian), 15, N 1: 107–111.
145. Sveshnikov BYa, Limareva LA (1960) On quenching of sensitized fluorescence in solutions. Dokl Akad Nauk USSR (in Russian), 133, N 4: 807–810.
146. Vekshin NL(1990) Exciplexes of pyrene-indole and pyrene-diethylaniline in liposomes and biomembranes. Zh Prikl Spektr (in Russian), 52, N 3: 471-476.
147. Forster Th (1959) Transfer mechanisms of electronic excitations. Disc Far Soc, N 27: 7–17.
148. Sveshnikov BYa, Kudryashov PI, Limareva LA (1960) On the sensitized fluorescence in solutions. Optika i Spektroskopiya (in Russian), 9, N 2: 203–208.

149. Forster Th (1949) Experimentelle und theoretische Untersuchung des zwischenmolekularen Ubergangs von Elektronenanregungsenergie. Z Naturforsch, 4a, H.5: 321–327.
150. Agranovich VM, Galanin MD (1978) Electronic Excitation Transfer in Condensated Media (in Russian). Moscow: Nauka.
151. Birks J, Munro I (1971) Fluorescence lifetimes of aromatic molecules. Uspekhi Fiz Nauk (in Russian), 105, N 2: 251–305.
152. F.R.G. patent No 2937352, class G01N21/01, 1981.
153. U.K. patent No 1593216, class G01N21/01, 1982.
154. U.S.A. patent No 4311387, class G01N21/64, 1983.
155. Vekshin NL (1986) Multipass luminescence cuvettes. In: Instruments and Laboratory Equipment for Research in New Trends of Biology and Biotechnology (in Russian). Pushchino; Acad Sci of the USSR, pp 34–37.
156. Vekshin NL (1987) Cuvettes of multiple reflection for luminescence analysis Instruments and Technique of Experiment (in Russian), N 1: 208–211.
157. Vekshin NL (1989) Multipass cuvettes for spectrofluorimetry. Anal Chim Acta, 227, N 1: 291–295.
158. Vekshin NL (1994) Multiple-pass cells for fluorescence spectroscopy. Optical Engineering Bull., N 3: 18–20.
159. Gilson TR, Hendra PJ (1973) Laser Raman Spectroscopy in Chemistry (in Russian). Moscow: Mir Publ.
160. O'Connor DV, Phillips D (1984) Time-correlated Single Photon Counting. N–Y: Acad.Press.
161. Eftink M (1990) Fluorescence techniques for studying protein structure. In: Protein Structure Determination, vol 35: Methods of Biochemical Analysis, C.H.Suelter (eds). John Wiley & Sons, Inc, pp 129–205.
162. Vekshin NL (1997) Energy Transfer in Macromolecules. SPIE, Bellmgham.
163. Demchenko AP (1986) Ultraviolet Spectroscopy of Proteins. Berlin: Springer.
164. Kronman MJ, Holmes LG (1971) The fluorescence of native denatured and reduced-denatured proteins. Photochem Photobiol, 14, N 2: 113–134.
165. Vekshin NL (1984) Luminescence studies on the conformational behavior of horse liver alcohol dehydrogenase. Eur J Biochem, 143, N 1: 69–72.
166. Brand L, Gohlike JR (1971) Nanosecond time-resolved fluorescence spectra of a protein-dye complex. J Biol Chem, 246, N 7: 2317–2324.
167. Vekshin NL (1987) On the changing of polarization degree across tryptophan fluorescence spectrum. Stud Biophys, 118, N 3: 173–182.
168. Irace G, Balestrieri C, Parlato G, Servillo L, Colonna G (1981) Tryptophanyl fluorescence heterogeneity of apomioglobins. Biochemistry, 20, N 4: 792–799.
169. Caswell DS, Spiro TG (1986) Tyrosine and tryptophan modification monitored by ultraviolet resonance Raman Spectroscopy. Biochim.Biophys.Acta, 873: 73–78.
170. Cho NJ, Song SH, Asher SA (1994) UV resonance Raman and excited-state relaxation rate studies of hemoglobin. Biochemistry, 33, N 19: 5932–5941.
171. Gorelik VS, Kozulin EA (1994) Two-photon-excited luminescence of biologically active solid state structures. Quantum Electronics, 24 (5): 462–463.
172. Rehms AA, Callis PR (1993) Two-photon fluorescence excitation spectra of aromatic amino acids. Chem Phys Lett, 208, N 3–4: 276–282.
173. Lloyd JBF, Evett IW (1977) Prediction of peak wavelengths and intensites in synchronously exited fluorescence emission spectra. Analyt Chem, 49, N 12: 1710–1715.

174. Romanovskaya GI, Pivovarov VM, Chibisov AK (1987) Use of synchronous spectrofluorimetry for the analysis of multicomponent mixtures. Zh Analiticheskoi Chimii (in Russian), 42, N 8: 1401–1406.
175. Romanovskaya GI, Chibisov AK (1987) Application of derivatives of synchronous luminescence spectra for qualitative analysis of mixtures of some aromatic substances. Zh Analiticheskoi Chimii (in Russian), 43, N6: 1120–1124.
176. Nekrasov VV, Volkova LV (1991) Selection of centers of specifical solvation of luminophores in solutions. Zh Prikladnoi Spectroscopil (in Russian), 55, N 5: 806–810.
177. Polyakov YaC, Shiffers LA (1984) Some new methods for fluorescence analysis of multicomponent mixtures. Zh Prikladnoi Spectroscopii (in Russian), 41, N 2: 181–190.
178. Stevenson OL, Johnson RW, Vodinh T (1994) Synchronous luminescence: A new detection technique for multiple fluorescent probes used for DNA sequencing. Biotechniques, 16, N 6: 1104–1110.
179. Miller JN (1979) Recent advances in molecular luminescence analysis. Proceed Analyt Divis Chem Soc. 16: 203–208.
180. Vekshin NL (1996) Division of tyrosine and tryptophan components by synchronous scanning method. Biofizika (in Russian), 41, N 6: 1176–1179.
181. Postnikova GB, Yumakova EM, Vekshin NL (1986) Study of conformational properties ofmyoglobin structure by the luminescence method. Biokhimiya (in Russian), 51, N 2: 313–320.
182. Calhoun DB, Vanderkooi JM, Englander SW (1983) Penetration of small molecules into proteins studied by quenching of phosphorescence and fluorescence. Biochemistry, 22: 1533–1539.
183. Vekshin NL (1987) Photoconformational relaxation of the protein structure determined by tryptophan fluorescence. Biofizika (in Russian), 32, N 4: 588–591.
184. Fleming GR, Morris JM, Robbins RJ, Woolfe GJ, Thistlethwaite PJ (1978) Nonexponential fluorescence decay of aqueous tryptophan and two related peptides by picosecond spectroscopy. Proc Natl Acad Sci USA. 75, N 10: 4652–4656.
185. Callis PR, Vivian JT, Slater LS (1995) Ab initio calculations of vibronic spectra for indole. Chem Phys Lett, 244: 53–58.
186. Valeur B, Weber G (1977) Resolution of the fluorescence excitation spectrum of indole into the lLa and ILb excitation bands. Photochem Photobiol, 25: 441–444.
187. Vivian JT, Callis PR (1994) Vibronic band shapes for indole from scaled bond order changes. Chem Phys Lett, 229: 153–160.
188. Burstein EA (1983) Intrinsic luminescence of light as a method for studying the fast structural dynamics. Molekularnaya Biologiya (in Russian), 17, N 3: 455–467.
189. Turoverov KK, Kuznetsova IM (1985) Polarization of intrinsic protein fluorescence; changes in polarization degree across the emission spectrum. Molekularnaya biologiya (in Russian), 19, N 5: 321–1331.
190. Mazurenko YuT, Bakhshiev NG, Piterskaya IV (1968) On the spectral dependences of the degree of rotational fluorescence depolarization of complex molecules in viscous solutions. Optika i Spektroskopiya (in Russian), 25, N 1: 92–97.
191. Lakowicz JR, Maliwal BP, Cherek H, Balter A (1983) Rotational freedom of tryptophan residues in proteins and peptides. Biochemistry, 22: 1741–1752.
192. Eftink M (1983) Quenching-resolved emission anisotropy studies with single and multitryptophan-containing proteins. Biophys J, 43: 323–334.
193. Mazurenko YnT, Udal'tsov VS (1982) Kinetic fluorescent spectroscopy of relaxation

processes in excited molecular systems. In: Excited Molecules, Kinetics of Transformations (in Russian). Leningrad: Nauka, pp 103–116.
194. Stepanov BI, Rubinov AN, Tomin VI (1982) Study of a dynamic heterogeneous orientational widening of electronic levels of dyes in solutions by the methods of high-resolution spectroscopy. Izv Akad Nauk USSR, Ser. Fizika (in Russian), 46, N 2: 380–387.
195. Rudik KI, Pikulik LG (1971) On the effect of exciting light on the fluorescence spectra of phthalanide solutions. Optika i Spectroskopiya (in Russian), N2: 275–278.
196. Mukherjee S, Chattopadhyay A (1994) Motionally restricted tryptophan environments at the peptide-lipid interface of gramicidin channels. Biochemistry, 33, N 17: 5089–5097.
197. Vanderkooi JM, Moy VT, Maniara G, Koloczek H, Paul KG (1985) Site-selected fluorescence spectra of porphyrin derivatives of heme proteins. Biochemistry, 24: 7931–7935.
198. Rubinov AN., Tomin VI (1983) Spectral properties of liquid polar solutions under conditions of dynamic heterogeneous orientational widening. Zh Prikl Spectres, (in Russian), 38, N 1: 42–61.
199. Gudgin-Templeton EF, Ware WR. The time dependence of the low-temperature fluorescence of tryptophan. J Phys Chem., 88, 4626–4631.
200. Lakowicz JR, Cherek H (1980) Dipolar relaxation in proteins on the nanosecond timescale observed by wavelength-resolved phase fluorimetry. J Biol Chem, 255, N 3: 831–834.
201. Meech SR, Phillips D (1982) Time-resolved emission spectroscopy of 1,3-dimethyl indole. Chem Phys Lett, 92: 523–526.
202. Cockle S.A., Szabo A.G (1981) Time-resolved fluorescence spectra of glucagon. Pholochem Photobiol., 34, N 1: 23-27.
203. Werner TC, Forster LC (1979) The fluorescence of tryptophyl peptides. Photochem Photobiol, 29, N 5: 905–914.
204. Harris DL., Hudson BS (1990) The photophysics of tryptophan in bacteriophage T4 lysozymes. Biochemistry, 29: 5276–5281.
205. Hudson B, Harris D (1990) T4 phage lysozyme: a protein designed for understanding tryptophan photophysics. In: Time-Resolved Laser Spectroscopy in Biochemistry, J.R. Lakowicz et al (eds).
206. Ichiye T, Karplus M (1983) Fluorescence depolarization of tryptophan residues in proteins; a molecular dynamics study. Biochemistry, 22, N 12: 2884–2893.
207. Munro I, Pecht I., Stryer L (1979) Subnanosecond motions of tryptophan residues in proteins. Proc. Natl. Acad. Sci, 76, N 1: 56–60.
208. Rubenchik AYa, Konev SV (1976) Some structural aspects of the interaction of yeast hexokinase with insulin. Molekularnaya Biologiya (in Russian), 10, N 1: 142–148.
209. Greed D (1984) The photophysics and photochemistry of the near-UV absorbing amino acids: tryptophan and its simple derivatives. Photochem Photobiol, 39, N 4: 537–562.
210. Konev SV, Volotovsky ID (1979) Effect of UV-light on proteins in solution and in biological membranes. In: Photobiology of the Animal Cell (in Russian). Leningrad: Nauka, pp 5–16.
211. Grossweiner LI (1976) Photochemical inactivation of enzymes. Current Topics in Rad Res, 11: 141–199.
212. Ostashevsky IYa, Zhuikov AG, Sezrakova AG (1979) Mechanisms of photoinactiva-

tion of chymotrypsin and trypsin. In: Photobiology of the Animal Cell (in Russian). Leningrad: Nauka, pp 40–42.
213. Wells TA, Nakazawa M, Manabe K, Song PS (1994) A conformational change associated with the phototransformation of *Pisum* phytochrome A as probed by fluorescence quenching. Biochemistry, 33, N 3: 708–712.
214. Mazul VM, Ermolaev YS, Konev SV (1980) Tryptophan phosphorescence at room temperature Zh Prikl Spektr. (in Russian), 32, N 5: 903–907.
215. Schlyer BD, Schauerle JA, Steel DG, Gafni A (1994) Time-resolved room temperature protein phosphorescence: nonexponential decay from single emitting tryptophans. Biophys J, 67, N 3: 1192–1202.
216. Golub AS, Popel AS, Zheng L, Pittman R.N (1997) Analysis of phosphorescence in heterogeneous systems using distributions of quencher concentration. Biophys J, 73: 452–465.
217. Strambini GB, Gonnelli M (1990) Tryptophan luminescence from liver alcohol dehydrogenase in its complexes with coenzyme. Biochemistry, 29: 196–203.
218. Sytnik A, Chumachenko YV, Demchenko AP (1991) Spectroscopic evidence for NADH-induced conformational changes in rabbit muscle aldolase. Biochim Biophys Acta, 1079: 123–127.
219. Vekshin NL, Vincent M, Gallay J (1992) Excited state lifetime distributions of tryptophan fluorescence in polar solutions: evidences for solvent exciplex formation. Chem Phys Lett, 199: 459–464.
220. Willis KJ, Neugebauer W, Sikorska M, Szabo AG (1994) Probing α-helical secondary structure at a specific site in model peptides via restriction of tryptophan side-chain rotamer conformation. Biophys J, 66, N 5: 1623–1630.
221. Chen Y, Gai F, Petrich JW (1994) Single-exponential fluorescence decay of the nonnatural amino-acid 7-azatryptophan and the nonexponential fluorescence decay of tryptophan in water. J Phys Chem, 98, N 8: 2203–2209.
222. Alcala JR, Gratton E, Prendergast FG (1987) Interpretation of fluorescence decays in proteins using continuous lifetime distributions. Biophys J, 51: 925–936.
223. Szabo AG, Rayner DM (1980) Fluorescence decay of tryptophan conformers in aqueous solution. J Am Chem Soc, 102, N 2: 554–563.
224. Beddard GS, Fleming GR, Porter SG, Robbins RS (1980) Time-resolved fluorescence from biological systems: tryptophan and peptides. Phil Trans R Soc, A 298, N 1439: 321–334.
225. Szabo AG, Rayner DM (1980) The time resolved emission spectra of peptide conformers measured by pulsed laser excitation. Biochem Biophys Res Com, 94, N 3: 909–915.
226. Arnold S, Tong L, Sulkes M (1994) Fluorescence lifetimes of substituted indoles in solution and in free jets: evidence for intramolecular charge-transfer quenching. J Phys Chem, 98, N 9: 2325–2327.
227. Hershberger MV, Lumry R, Verall R (1981) The 3-methylmdol-ri-butanol exciplex sites in indole compounds. Photochem Photobiol, 33, N 5: 609–617.
228. Vincent M, Gallay J, Demchenko AP (1995) Solvent relaxation around the excited state of indole: analysis of fluorescence lifetime distributions and time-dependence spectral shifts. J Phys Chem, 99: 14931–14941.
229. Vincent M, Broclion JC, Merola F, Jordi W, Gallay J (1988) Nanosecond dynamics of horse heart apocytochrome c in aqueous solution as studied by time-resolved fluorescence of the single tryptophan residue (Trp-59). Biochemistry, 27: 8752–8761.

230. Sopkova J, Gallay J, Vincent M, Pancoska P, Lewitbentley A (1994) The dynamic behavior of annexin-V as a function of calcium ion binding. A circular-dichroism, UV absorption, and steady-state and time-resolved fluorescence study. Biochemistry, 33, N 15: 4490–4499.
231. Sopkova J, Gallay J, Vincent M, Pancoska P, Lewit-Bentley A (1994) The dynamics behaviour of annexin V as a function of calcium ion binding. Biochemistry, 33: 4490–4499.
232. Grinvald A, Steinberg IZ (1974) Fast relaxation processes in a protein revealed by the decay kinetics of tryptophan fluorescence. Biochemistry,13: 5170–5178.
233. Wolfbeis OS (1985) The fluorescence of organic natural products. In: Molecular Luminescence Spectroscopy, S.Schulman (eds), vol 1, Wiley, N–Y, pp 167–369.
234. Muino PL, Callis PR (1994) Simulations of solvent effects on fluorescence spectra and dynamics of indoles. In: Time-Resolved Laser Spectroscopy in Biochemistry IY, 2137: 362–371.
235. Muino PL, Callis PR (1994) Hybrid simulations of solvation effects on electronic spectra: indoles in water. J. Chem. Phys., 100, N 6: 4093–4109.
236. Szymanski HA (1967) A Systematic Approach to the Interpretation of Infrared Spectra. Buffalo: Hertillon Press.
237. Kazitsina LA, Kupletskaya NB (1979) Application of UV-, IR-, NMR- and Mass-spectroscopy in Organic Chemistry (in Russian). Moscow: Moscow State University.
238. Vekshin NL (1991) Excitation conversion in excimers and exciplexes of aromatic hydrocarbons. Spectrochim Acta, 47A, N 1: 155–158.
239. Vekshin NL (1989) Fractional energy transfer in excimers and exciplexes. Preprint. Pushchino: Acad Sci USSR, pp 1–25.
240. Huang Y, Sulkes M (1996) Anomalously short fluorescence lifetime in jet cooled 4-hydroxyindole. Chem Phys Lett, 254: 242–248.
241. Boens N, Janssens LD, Dommelen LV, De Schryver PO, Gallay J (1992) Photophysics of tryptophan: global analysis of the fluorescence decay surface as a function of pH, temperature, quencher concentration, excitation and emission wavelengths, timing calibration and deuterium isotope effect. In: Time-Resolved Laser Spectroscopy in Biochemistry III, 1640: 58–69.
242. Longworth JW, Battista MDC (1970) Exciplexes and the fluorescence of Staphylococcus aureus endonuclease. Photochem Photobiol, 12, N 1: 29–35.
243. Szabo AG, Stepanik TM, Wayner DM, Yoyng NM (1983) Conformational heterogenety of the copper binding site in azurin. Biophys J, 41: 233–244.
244. Vigny P, Duquesne M (1976) On the fluorescence properties of nucleotides and polynucleotides at room temperature. In: Excited States of Biological Molecules, J.B.Birks (eds). London: John Wiley-Sons, pp 167–177.
245. Helene C (1976) Excited state interactions and energy transfer processes in the photochemistry of protein-nucleic acid complexes In: Excited States of Biological Molecules, J.B. Birks (eds). London. John Wiley-Sons, pp 151–166.
246. Doring K, Beck W, Konermann L, Jahnig F (1997) The use of a long-lifetime component of tryptophan to detect slow orientational fluctuations of proteins. Biophys J, 72: 326–334.
247. Gilardi G, Mei G, Rosato N, Canters GW, Finazzi-Agro A (1994) Unique environment of Trp48 in Pseudomonas Aeruginosa azurin as probed by site-directed mutagenesis and dynamic fluorescence spectroscopy. Biochemistry, 33, N 6: 1425–1432.
248. Bastiaens PIH, Van Hoek A, Wolkers WF, Brochon JC, Visser AJWG (1992) Dy-

namics and geometry in comparison of the dynamical structure of lipoamide dehydrogenase and glutathione reductase by time-resolved polarized flavin fluorescence. Biochemistry, 31, N 31: 7050–7060.
249. Barashkov NN, Sakhno TV, Nurmukhametov RN, Khahel' OA (1993) Excimers of organic molecules. Uspekhi Khimii (in Russian), 62 (6): 579–593.
250. Forster Th (1975) Excimers and exciplexes. In: The Exciplex. N–Y : Acad Press, pp 1–15.
251. De Schryver PO, Declercq D, Depaemelaere S, Hermans E, Onkelinx A, Verhoeven JW, Gelan J (1994) Photophysics of linked donor-acceptor systems, through-space and through-bond interactions J Photochem Photobiol A: Chem, 82: 171–179.
252. Anner O, Haas Y (1985) Jet-cooled exciplexes: dimethylaniline with anthracene and with perylene. Chem Phys Lett, 119, N 2–3: 199–205.
253. Schneider S, Rehaber H, Gahr M, Schublbauer W (1994) Photophysical and photochemical behaviour of (N, N-dimethylamino)styrenes. J Photochem Photobiol A: Chem, 82: 129–136.
254. Brancaleon L, Crippa PR, Minari C (1994) Time-resolved photoacoustic calorimetry of tryptophan in water and organic solvents. Photochem Photobiol, 59, N 6: 590–595.
255. Muino PL, Harris D, Berryhill J, Hudson B, Callis PR (1992) Simulations of solvent dynamics effects on the fluorescence of 3-methyhndole in water. In: Time-Resolved Laser Spectroscopy in Biochemistry III, 1640: 240–251.
256. Callear AB, Lambert JD (1973) Energy exchange between chemical particles. In: Exciter Particles in Chemical Kinetics (in Russian). Moscow: Mir, pp 214–317.
257. Medvedev ES, Osherov VI (1983) Theory of Nonradiative Transitions in Polyatomic Molecules (in Russian). Moscow: Nauka.
258. Tiang Ch-H (1969) Theory on the radiationless transitions in large polyatomic molecules. Photochem Photobiol, 9, N 1: 17–31.
259. Robl T, Seilmeier A (1988) Ground-state recovery of electronically excited malachite green via transient vibrational heating. Chem Phys Lett, 147, N 6: 544–550.
260. Dexter DL (1973) Novel investigations of sensitized luminescence and energy transfer in solids. Izv Acad Sci of the USSR, ser. Physics, (in Russian), 37, N 2: 257–261.
261. Dong DC, Winnik MA (1982) The Py scale of solvent polarities. Solvent effects on the vibronic fine structure of pyrene fluorescence and empirical correlations. Photochem Photobiol, 35, N 1: 17–21.
262. Kalyanasundaram K, Thomas JK (1977) Environmental effects on vibronic band intensities in pyrene monomer fluorescence and their application in studies of micellar systems. J Am Chem Soc, 99, N 7: 2039–2045.
263. Vekshin NL (1989) Exciplexes of pyrene with indole and diethylaniline in liposomes and biomembranes Analyt Chim Acta, 227, N 1: 267–272.
264. Ruttens F, Goedeweeck R, Lopez-Arbeloa F, De Schryver FC (1985) Intramolecular exciplex formation in N-acetyl-1-pyrenyl-alanyl- 1- methyltryptophan methylester. Photochem Photobiol, , 42, N 4: 341–346.
265. Forster Th (1948) Zwischenmolekulare energiewanderung und fluoreszenz. Ann.Phys., 2, N 1–2: 55–75.
266. Dexter DL (1953) A theory of sensitized luminescence in solids. J Chem Phys, 21: 836–850.
267. Kaplan IG (1982) Introduction in Theory of Intermolecular Interactions (in Russian). Moscow: Nauka.
268. Galanin MD (1955) On the effect of concentration on luminescence of solution. Zh

Ekspnr i Teor Fiziki(in Russian),28, N 4: 485–495.
269. Ermolayev VL, Bodunov EN, Sweshnikova EB, Shakhverdov TA.(1977). Nonradiative Transfer of Electronic Excitation Energy (in Russian). Leningrad: Nauka.
270. Chernyakovsky FP (1983) Application of electrochromism for studies of bio- and synthetic polymers. In: Itogi nauki i techniki, ser. Biofizika (in Russian), vol 15, Moscow: VINITI.
271. Shahparonov MI (1983) Intermolecular Interactions (in Russian). Moscow: Znanie.
272. Gaisenok VA, Sarzhevsky AM (1986) Anisotropy of Absorption and Emission of Multiatom Molecules (in Russian). Minsk University.
273. Stryer L, Haugland RP (1967) Energy transfer: a spectroscopic ruler. Proc Natl Acad Sci USA, 58, N 2: 719–726.
274. Vekshin NL (1987) Transfer of electronic excitation energy in solutions of organic compounds (in Russian). Preprint. Pushchino: Acad Sci USSR.
275. Bravo J, Mendicuti F, Saiz E, Mattice WL (1994) Intramolecular energy transfer in bichromophoric diesters containing anthracene and naphthalene groups separated by different numbers of methylene spacers. Macromol Chem Phys, 195: 3411–3424.
276. Knox RS, Gulen D (1993) Theory of polarized fluorescence from molecular pairs: Forster transfer at large electronic coupling. Photochem Photobiol, 57, N 1: 40–43.
277. Maier GV, Kopylova TN, Artyukhov VY, Samsonova LG, Karypov AV, Rib NR, Park GD (1993) Effects of intramolecular energy transfer on spectro-luminescent and lasing properties of bifluorophores molecules. Optika i Spektr. (in Russian), 75, N 2: 337–343.
278. Reimers JR, Hush NS (1994) Electron transfer and energy transfer through briged systems. Tight-binding linkages with zero or non-zero asymptotic band gap. J Photochem Photobiol, B: Biol, 82: 31–46.
279. Finckh P, Heitele H, Michel-Beyerle ME (1989) Intramolecular electron transfer in viscous solution. Chem Phys, 138: 1–10.
280. Heitele H, Finckh P, Weeren S, Pollinger F, Michel-Beyerle ME (1989) Solvent polarity effects on intramolecular electron transfer: energetic aspects. J Phys Chem, 93: 5173–5179.
281. Tamiaki H, Nomura K, Maruyama K (1994) Energy and Electron-transfer in synthetic oligoproline-bridged porphyrin donor-acceptor molecules. Bull Chem Soc Japan, 67, N 7: 1863–1871.
282. Zenkevich EI, Chernook AV, Shulga AM, Pershukevitch PP, Gurinovitch, Sagun EI (1991) Role of dipole-dipole and exchange transfer processes in the electron excitation deactivation in chemical dimers of cyclopentanporphyrines. Chim Phys (in Russian), 10, N9: 1183–1191.
283. Heitele H, Pollinger F, Haberle T, Michelbeyerle ME, Staab HA (1994) Energy-gap and temperature-dependence of photoinduced electron transfer in porphyrin-quinone cyclophanes. J Phys Chem., 98, N 30: 7402–7410.
284. Indelli MT, Bignozzi CA, Harriman A, Schoonover JR, Scandola F (1994) Four intercomponent processes in a Ru(II)-Rh(III) polypyridine dyad: electron transfer from excited donor, electron transfer to excited acceptor, charge recombination, and electronic energy transfer. J Am Chem Soc., 116, N 9: 3768–3779.
285. Spooner SP, Whitten DG (1994) Energy and electron transfer processes through Langmuir-Blodgett multilayers formed from diphenylpolyene surfactants. J Am Chem Soc, 116, N 4: 1240–1248.
286. Oevering H, Verhoeven JW, Paddon-Row MN, Warman JM (1989) Charge-transfer

absorption and emission resulting from long-range through-bond interaction: exploring the relation between electronic coupling and electron-transfer in briged donor-acceptor systems. Tetrahedron, 45, N 15: 4751–4766.
287. Clayton AHA, Ghiggino KP, Lawson JM, Paddonrow MN (1994) Electron and singlet energy transfer in rigid supramolecular systems. J Photochem Photobiol, A: Chem, 80, N 1–3: 323–331.
288. Tehver IY, Hizhnyakov VV (1975) Nonradiative transfer of electronic excitation during vibrational relaxation. Zh Experim Theor Fiziki (in Russian), 69, N 2: 599–610.
289. Valeur B, Weber G (1978) A new red-edge effect in aromatic molecules: anomaly of apparent rotation revealed by fluorescence polarization. J Chem Phys, 69: 2393–2400.
290. Wilson R, Callis PR (1976) Excitons, energy transfer, and charge resonance in excited dinucleotides and polynucleotides. A photoselection study J Phys Chem, 80: 2280–2288.
291. Wu PG., Fujimoto BS, Song L, Schurr JM (1991) Effect of ethydium on the torsion constants of linear and supercoiled DNAs. Biophys Chem, 41: 217–236.
292. Rayner DM, Szabo AG, Loutfy RO, Yip RW (1980) Singlet energy transfer between nucleic acid bases and dyes in intercalation complexes. J Phys Chem, 84: 289–293.
293. Rubin YuV, Egupov SA (1987) Influence of intermolecular interactions on electron-excited states of nucleic acid components. Biofizika (in Russian), 32: 378–382.
294. Rapoport VL, Bakulev VM (1984) Spectral appearance of conformation heterogenity in cytidine nucleotide: fraction with strong exciton interaction between bases. Molecularnaya Biologiya (in Russian), 18: 382–389.
295. Rapoport VL, Kononov AI (1988) Luminescent conformers of adenine dinucleotide with strong exciton interaction between nitrogen bases. Doklady Akademn Nauk (in Russian), 298: 231–235.
296. Tehver IYu, Hizhnyakov VV (1974) Transfer of electronic excitation during vibrational relaxation Letters to Zh Experim Theor Fiziki (in Russian), 19, N 6: 338–342.
297. Razzhivin AP, Ledneva RK, Terganova GV, Borisov AY, Bogdanov AA, Kost AA (1979) Interaction of tabac mosaic virus protein with synthetic polynucleotides, contained fluorescent lable: absorption and fluorescence of copolymers of adenylic and ethenoadenylic acids. Bioorganicheskaya Khimiya (in Russian), 5: 691–700.
298. Dalgleish DG, Peacocke AR, Fey G, Harvey C (1971) The circular dichroism in the ultraviolet of aminoacridines and ethidium bromide bound to DNA. Biopolymers, 10: 1853–1863.
299. Poletaev AI (1976) Circular dichroism of DNA complexes. In: Molekularnaya biologiya (in Russian), vol 8, part 2. Moscow: VINITI, pp 180–241.
300. Nucleic Acid Geometry and Dynamics (1980) Sarma RH (eds). Pergamon Press, N-Y.
301. Paoletti J, Le Pecq JB (1971) Resonance energy transfer between ethidium bromide molecules bound to nucleic acids; does intercalation wind or unwind the DNA helix? J Mol Biol, 59: 43–62.
302. Genest D, Wahl Ph (1978) Pulse fluorimerty study in polarized light of DNA-ethidium bromide complexes. Biophys Chem, 7: 317–323.
303. Van Nostrand F, Pearlstein RM (1976) Singlet excitation transfer from poly rA to bound dye at 77 K. Chem Phys Lett, 39: 269–272.
304. Kumar CV, Asuncion EH (1994) DNA binding studies and site selective fluorescence

sensitization of an anthryl probe. J Am Chem Soc, 116, N 1: 403–404.
305. Kumar CV, Chaudhari A (1994) Enhanced energy transfer between aromatic chromophores bound to hydrophobically-modified layered zirconium-phosphate suspensions. J Am Chem Soc, 116, N 1: 403–404.
306. Selvin PR, Rana TM, Hearst JE (1994) Luminescence resonance energy transfer. J Am Chem Soc, 116, N 13: 6029–6030.
307. Clegg RM, Murchie AIH, Zechel A, Lilley DMJ (1993) Observing the helical geometry of double-stranded DNA in solution by fluorescence resonance energy transfer. Proc. Natl. Acad. Sci., 90, N 7: 2994–2998.
308. Parkhurst KM, Parkhurst LJ (1996) Detection of point mutations in DNA by fluorescence energy transfer. J Biomed Optics, 1 (4): 435–441.
309. Ha T, Enderle Th, Ogletree DF, Chemla DS, Selvin PR, Weiss S (1996) Probing the interaction between two single molecules: fluorescence resonance energy transfer between a single donor and a single acceptor. Proc Natl Acad Sci USA, 93: 6264–6268.
310. Ha T, Enderle Th, Chemla DS, Selvin PR, Weiss S(1996) Single molecular dynamics studied by polarization modulation. Phys Rev Lett, 77, N 19: 3979–3982.
311. Sixou S, Szoka FC, Green GA., Giusti B, Zon G, Chin DJ (1994) Intracellular oligonucleotide hybridization detected by fluorescence resonance energy transfer. Nucleic Acids Research, 22, N 4: 662–668.
312. Moreno MJ, Prieto M (1993) Interaction of the peptide hormone adrenocorticotropin, ACTH(1–24), with a membrane model system: a fluorescence study. Photochem Photobiol, 57, N 3, 431–437.
313. Burstein EA (1976) Luminescence of protein chromophores. In.: Itogi Nauki i Techniki, ser. Biofizika (in Russian), vol 6. Moscow: VINITI, pp 16–120
314. Borkman RE, Puillips SR (1985) Tyrosine-to-tryptophan energy transfer and the structure of calf gamma crystallin. Exp Eye Res, 40: 819–826.
315. Lerner J, Lami H(1976) Electronic energy transfer in some class B proteins: trypsin, Lysozyme, chymotrypsin and chymotrypsinogene A. In: Excited States of Biological Molecules. London: John Wiley-Sons, 601–611.
316. Saito Y, Tachibana H, Hayashi H, Wada A (1981) Excitation-energy transfer between tyrosine and tryptophan in proteins. Phofochem Photobiol, 33, N 3: 289–295.
317. Turoverov KK, Kuznetsova IM, Zaitsev VN (1984) Interpretation of UV-fluorescence of azurin on X-ray analysis data. Bioorganicheskaya Khimiya (in Russian), 10: 792–806.
318. Wu PG, Li YK, Talalay P, Brand L (1994) Characterization of the 3 tyrosine residues of delta(5)-3-ketosteroid isomerase by time-resolved fluorescence and circular dichroism. Biochemistry, 33, N 23: 7415–7422.
319. Desie G, Boens N, De Schryver FC (1986) Study of the time-resolved tryptophan fluorescence of crystalline chymotrypsin. Biochemistry, 25: 8301–8308.
320. Luisi PL, Favilla R (1970) Tryptophan fluorescence quenching in horse liver alcohol dehydrogenase. Eur J Biochem, 17, N 1: 91–94.
321. Knutson JR, Walbridge DG, Brand L (1982) Decay-associated fluorescence spectra and the heterogeneous emission of alcohol dehydrogenase. Biochemistry, 21: 4671–4675.
322. Hardman M (1985) Carboxymethylated liver alcohol dehydrogenase: pH dependence of hydride transfer during ethanol oxidation. Biochim Biophys Acta, 831: 347–349.
323. Vekshin NL (1985) Luminescence study of the conformational behavior of horse

liver alcohol dehydrogenase upon substrate binding Molekularnaya Biologiya (in Russian), 19, N 3: 767–773.
324. Ross JBA, Schmidt CJ, Brand L (1981) Time-resolved fluorescence of two tryptophans in horse liver alcohol dehydrogenase. Biochemistry, 20, N 15: 4369–4377.
325. Schellenberg KA (1965) Evidence for the participation of tryptophan as intermediate in transfer of hydrogen between NADH and substrate. J Biol Chem, 240: 1165–1170.
326. Boyer PD (1974) Conformational coupling in biological energy transductions. In: Dynamics of Energy-Transducing Membranes. Ernster L (eds). Amsterdam: Elsevier, pp 289–301.
327. Cedergren-Zeppezauer ES, Andersson I, Ottonello S, Bignetti E (1983) X-ray analysis of structural changes induced by NADH when bound to alcohol dehydrogenase. Biochemistry, 24, N 15: 4000–4010.
328. Sebban P, Coppey M, Alpert B, Lindqvist L, Jameson DM (1980) Fluorescence properties of porphyrin-globin from human hemoglobin Photochem Photobiol, 32: 727–731.
329. Vekshin NL (1983) Checking of Forster's concepts of the inductive resonance energy transfer for the tryptophan-pyrene pair. Molekularnaya Biologiya (in Russian), 17, N 4: 827–832.
330. Vauhkonen M, Somerharju P (1989) Parinaroyl and pyrenyl phospholipids as probes for the lipid surface layer of human low density lipoproteins. Biochim Biophys Acta, 984: 81–87.
331. Khahel' OA, Nurmukhametov RN, Sakhno TV, Serov SA, Barashkov NN, Murav'eva TM (1992) Dimers of pyrene in a polymer matrice Zh Phys Chimii (in Russian), 66, N 10: 2639–2645.
332. Stryer L (1978) Fluorescence energy transfer as a spectroscopic ruler. Ann Rev Biochem, 47: 819–846.
333. Alfimova EYa, Likhtenstein GI (1976) Fluorescence study of energy transfer as method of study of protein structure. In: Molecular Biology (in Russian), vol 8, part 2. Moscow: VINITI, pp 127–179.
334. Wu P, Brand L (1994) Resonance energy transfer: methods and applications. Analyt Biochem, 218, N 1: 1–13.
335. Graceffa P (1997) Arrangement of COOH-terminal and NH_2-terminal domains of caldesmon bound to actin. Biochemistry, 36: 3792–3801.
336. Steer BA, Merrill AR (1994) The colicin El insertion competent state. Detection of structural changes using fluorescence resonance energy transfer. Biochemistry, 33, N 5: 1108–1115.
337. Houtzager V, Ouellet M, Falgueyret JP, Passmore LA, Bayly C, Percival MD (1996) Inhibitor-induced changes in the intrinsic fluorescence of human cyclooxygenase-2. Biochemistry, 35: 10974–10984.
338. Gabellieri E, Strambini GB, Baracca A, Solaini G (1997) Structural mapping of the e-subunit of mitochondrial H^+-ATPase complex (Fl). Biophys J, 72: 1818–1827.
339. Vekshin NL, Sukharev VI, van Hoek A, Visser AJWG (1999) Competition between energy transfer and deactivation during quenching of tryptophan fluorescence of albumin by dye molecules. J Fluorescence. 9, N 2: 99–101.
340. Lakey JH, Duche D, Gonzalez-Manas JM, Baty D, Pattus F(1993) Fluorescence energy transfer distance measurements. The hydrophobic helical hairpin of colicin A in the membrane bound state. J Mol Biol, 230: 1055–1067.
341. Haynes MP, Chong PLG, Buckley HR, Pieringer RA (1996) Fluorescence studies on

the molecular action of amphotericine B on susceptible and resistant fungal cells. Biochemistry, 35: 7983–7992.
342. Schumann M, Dathe M, Wieprecht T, Beyermann M, Bienert M (1997) The tendency of magainin to associate upon binding to phospholipid bilayers. Biochemistry, 36: 4345–4351.
343. Moll G, Papini E, Colonna R, Burroni D, Telford J, Rappuoli R, Montecucco C (1995) Lipid interaction of the 37-kDa and 58-kDa fragments of the Helicobacter pilori cytotoxin. Eur J Biochem, 234: 947–952.
344. Runnels LW, Scarlata SF (1995) Theory and application of fluorescence homotransfer to melittin oligomenzation. Biophys J, 69: 1569–1583
345. Adair BD, Engelman DM (1994) Glycophorin A helical transmembrane domains dimerize in phospholipid bilayers. A resonance energy-transfer study. Biochemistry, 33, N 18: 5539–5544.
346. Abrams FS, London E (1993) Extension of the parallax analysis of membrane penetration depth to the polar region of model membranes: use of fluorescence quenching by a spin-label attached to the phospholipid polar headgroup. Biochemistry, 32: 10826–10831.
347. Yao Y, Gao J, Squier TC (1996) Dynamic structure of the calmodulin-binding domain of the plasma membrane Ca-ATPase in native erythrocyte ghost membranes. Biochemistry, 35: 12015–12028.
348. Shinitzky M, Rivnay B (1977) Degree of exposure of membrane proteins determined by fluorescence quenching. Biochemistry, 16, N 5: 982–986.
349. Pechatnikov VA., Ivkova MN, Rizvanov FF, Pletnev VV (1979) Formation of transmembrane potential upon ATP hydrolysis in the sarcoplasmic reticulum. Biofizika (in Russian), 24, N 3: 476–483.
350. Kotelnikova RA, Tatyanenko LV, Moshkovsky YuSh (1987) Localization of tryptophan residues in Ca^{2+}-ATPase of the sarcoplasmic reticulum. Biofizika (in Russian), 32, N 3: 420–423.
351. Augustin J, Hasselbach W (1973) Studies on the fluorescence of ANS by the membranes of the sarcoplasmic reticulum. Eur.J.Biochem., 35, N 1: 114–121.
352. Chernogryadskaya NA, Rozanov YuM, Bogdanova MS, Borovikov YuS (1978) Ultraviolet Fluorescence of Cell (in Russian). Leningrad, Nauka.
353. Katsumata C, Miyazi A, Ozawa T (1970) Fluorescence changes in mitochondrial protein associated with energy state J Biochem, 68, N 3: 423–435.
354. Sorokovoy VI, Klebanov GI, Vladimirov YuA (1972) On the nature of protein fluorescence in the mitochondria in different functional states Molek Biol (in Russian), 6, N 5: 705–711.
355. Vekshin NL (1998) Resonant energy transfer from proteins to pyridine nucleotides in mitochondria. Biochemistry (Moscow). vol 63, N 9: 1110–1113.
356. Sundstrom V, Grondell Van R, Bergstrom H, Akesson E, Gillbro T (1986) Excitation-energy transport in the bacteriochlorophyll antenna systems of Rhodospirillum Rubrum and Rhodobacter Sphaeroides studied by low-intensity picosecond absorption spectroscopy. Biochim Biophys Acta, 851: 431–446.
357. Wendler J, John W, Scheer H, Holzwarth AR (1986) Energy transfer in trimeric C-phycocyanin studies by picosecond fluorescence kinetics. Photochem Photobiol, 44: 79–85.
358. Dagen AJ, Alfano RR, Zilinskas BA, Swenberg CE (1986) Analysis of fluorescence kinetics and energy transfer in isolated subunits of phycoerythrin from Nostoc Sp.

Photochem Photobiol, 43: 71–79.
359. Sauer K, Scheer H, Sauer P (1987) Forster transfer calculation based on crystal structure data from Agmenellum Quadruplicatum C-phycocyanin. Photochem Photobiol, 46: 427–440.
360. MacColl R, Guard-Friar D, Ryan TJ, Csatorday K, Wu P (1988) The route of exciton migration in phycocyanin 612. Biochim Biophys Acta., 934: 275–281.
361. Guard-Friar D, MacColl R, Berns DS, Wittmershaus B, Knox RS (1985) Picosecond fluorescence of Cryptomonad biliproteins. Effect of the excitation intensity and the fluorescence decay times of phycocyanin 612, phycocyanin 645 and phycoerythrin 545. Biophys J, 47: 787-793.
362. Vasil'ev SS, Delver EP, Pashchenko VZ (1987) On question about energy transfer of electronic excitation in chromatophores of purpure bacteria. Biofizika (in Russian), 32: 263–268.
363. Huber R (1990) A structural basis of light energy and electron transfer in biology. In series: «Chemistry», N 1 (in Russian). Moscow: Znanie, pp 3–32.
364. Hess S, Visscher KJ, Pullerits T, Sundstrom V, Fowler GJS, Hunter CN (1994) Enhanced rates of subpicosecond energy-transfer in blue-shifted light-harvesting Lh2 mutants of Rhodobacter Sphaeroides. Biochemistry, 33, N 27: 8300–8305.
365. Photosynthesis; Chemical Models and Mechanisms, V.M.Cherkasov (eds), in Russian). Kiev: Naukova dumka, 1989.
366. Van der Meer BW (1988) Biomembrane structure and dynamics viewed by fluorescence. In.: Subcellular Biochemistry, v.13: Fluorescence Studies on Biological Membranes (eds by H.J.Hilderson and J.R.Harris). Plenum Press: N–Y, pp 1–53.
367. Haugland RP (1992) Handbook of Fluorescent Probes and Research Chemicals. USA: Mol Probs, Inc.
368. Lin T-I, Dowben RM (1982) Fluorescence spectroscopic studies of pyrene-actin adducts. BiophysChem, 15, N 4: 289–298.
369. Vanderkooi JM, Callis JB (1974) Pyrene: A probe of lateral diffusion in the hydrophobic region of membranes. Biochemistry, 13, N 19: 4000–4006.
370. Galla H.J., Sackmann E (1974) Lateral diffusion in the hydrophobic region of membranes: use of pyrene excimers as optical probes. Biochim Biophys Acta, 339, N 1: l03–115.
371. Fischkoff S, Vanderkooi JM (1975) Oxygen diffusion in biological and artificial membranes determined by the fluorochrome pyrene. J Gen Physiol, 65, N5: 663–676.
372. Cheng S, Thomas JK, Kulpa CF (1974) Dynamics of pyrene fluorescence in E.coli membrane vesicles. Biochemistry, 13, N 6: 1135–1139.
373. Vekshin NL (1987) On measuring biomembrane microviscosity using pyrene luminescence in aerobic conditions. J Biochem Biophys Methods, 15: 97–104.
374. L'Heureux GP, Fragata M (1987) Use of pyrene and 16(l-pyrenyl)-hexadecanoic acid incorporated in unilamellar phosphohpid vesicles as polarity probes of the hydrocarbon core. Biophys J, 51, N 2, pt.2, pp 531a.
375. Kinnunen PKJ, Koiv A, Mustonen P (1993) Pyrene-labelled lipids as fluorescent probes in studies on biomembranes and membrane models. In: Fluorescence Spectroscopy: New Methods and Applications, Wolfbeis O.S. (eds), Springer-Verlag, Berlin, pp 159–171.
376. Martins J, Vaz WLC, Melo E (1996) Long-range diffusion coefficients in two-dimensional fluid media measured by the pyrene excimer reaction. J Phys Chem, 100, N 5: 1889–1895.

377. Sassaroli M, Ruonala M, Virtanen J, Vauhkonen M, Somerharju P (1995) Transversal distribution of acyl-linked pyrene moieties in liquid-crystalline phosphatidylcholine bilayers: a fluorescence quenching study. Biochemistry, 34: 8843–8851.
378. Somerharju P, van Paridon PA, Wirtz KWA (1990) Application of fluorescent phospholipid analogues to studies on phospholipid transfer proteins. In: Subcellular Biochemistry, H. Hilderson (eds), 1990, Plenum Publ Corp pp 21–43.
379. Kierzek R, Li Y, Turner DH, Bevilacqua PC (1993). 5'-amino pyrene provides a sensitive, nonperturbing fluorescent probe of RNA secondary and tertiary structure formation. J Am Chem Soc, 115, N 12: 4985–4992.
380. Bevilacqua PC, Kierzek R, Johnson KA, Turner DH (1992) Dynamics of ribozyme binding of substrate revealed by fluorescence-detected stopped-flow methods. Science, 258: 1355–1358.
381. Hiratsuka T (1997) Monitoring the myosin ATPase reaction using a sensitive fluorescent probe: pyrene-labeled ATP. Biophys J, 72: 843–849.
382. Merzlyak MN, Kashulin PA (1991) Pyrene fluorescence probing of unsaturated lipids in *Phytophthora infestans* zoospores. Gen Physiol Biophys, 10, N 6: 561–568.
383. Lakowicz JR, Szmacinski H, Nowaczyk K, Berndt KW, Johnson ML (1993) Fluorescence lifetime imaging and application to Ca^{2+} imaging. In: Fluorescence Spectroscopy: New Methods and Applications, Wolfbeis O.S. (eds), Springer-Verlag, Berlin, pp 129–147.
384. Hirshfield KM, Toptygin D, Packard BS, Brand L (1993) Dynamic fluorescence measurements of two-state systems: application to calcium-chelating probes. Anal Biochem, 209: 209–218.
385. Rettig W (1993) Kinetic studies on fluorescence probes using synchrotron radiation. In: Fluorescence Spectroscopy: New Methods and Applications, Wolfbeis O.S. (eds), Springer-Verlag, Berlin, pp 31–47.
386. Lisi A, Pozzi D, Grimaldi S (1993) Use of the fluorescent probe laurdan to investigate structural organization of the vesicular stomatitis virus membrane. Membrane Biochemistry, 10, N 4: 203–212.
387. Grzesiek S, Otto H, Dencher NA (1989) pH-induced fluorescence quenching of 9-ammoacridine in lipid vesicles is due to excimer formation at the membrane. Biophys J., 55: 1101–1109.
388. Small JR, Larson SL (1990) Photoacoustic determination of fluorescent quantum yields of protein probes. In: Time-Resolved Laser Spectroscopy in Biochemistry II, J.R.Lakowicz (eds), SPIE Proceedings, vol 1209.
389. Crippa PR, Vecli A, Viappiani C (1994) Time-resolved photoacoustic spectroscopy: new developments of an old idea. J Photochem Photobiol, B Biol, 24, N 1: 3–15.
390. Leeson DT, Wiersma DA (1995) Real time observation of low-temperature protein motions. Phys Rev Lett, 74, N 11: 2138–2141.
391. Leeson DT, Wiersma DA (1995) Looking into the energy landscape of myoglobin. Nature Struct Biol, vol 2, N 10: 848–851.
392. Leeson DT, Berg O, Wiersma DA (1994) Low-temperature protein dynamics studied by the long-lived stimulated photon echo. J Phys Chem, 98: 3913–3916.
393. Rahman NA, Pecht I, Roess DA, Barisas BG (1992) Rotational dynamics of type I Fc receptors in individually-selected rat mast cells studied by polarized fluorescence depletion. Biophys J, 61, N 2: 334–346.
394. Young RM, Arnette JK, Roess DA, Barisas BG (1994) Quantitation of fluorescence energy transfer between cell surface proteins via fluorescence donor photobleaching

kinetics. Biophys J, 67, N 2: 881–888.
395. Huang Z, Pearce KH, Thompson NL (1994) Translation diffusion of bovine prothrombin fragment 1 weakly bound to supported planar membranes: measurement by total internal reflection with fluorescence pattern photobleaching recovery. Biophys J, 67, N 4: 1754–1766.
396. Thompson NL, Pearce KH, Hsieh HV (1993) Total internal reflection fluorescence microscopy: application to substrate-supported planar membranes. Eur Biophys .J, 22: 367–378.
397. Ladokhin AS, Selsted M, White SH (1997) Sizing membrane pores in lipid vesicles by leakage of co-encapsulated markers: pore formation by melittin. Biophys J, 72: 1762–1766.
398. Coelho FP, Vaz WLC, Melo E 91997) Phase topology and percolation in two-component lipid bilayers: a Monte Carlo approach. Biophys J, 72: 1501–1511.
399. Anderson S.R., Weber G (1969) Fluorescence polarization of the complexes of l-anilino-8-naphthalenesulfonate with bovine serum albumin. Biochemistry, 8: 371-377.
400. Vanderkooi J, Martonosi A (1971) Use of 8-anilino-l-naphthalene sulfonate as conformational probe on biological membranes. In: Probes of Structure and Function of Macromolecules and Membranes, vol. 1 B.Chance (eds), Acad Press, N–Y, 293–301.
401. Hardy, SJS. & Randall LL (1991) A kinetic partitioning model of selective binding of nonnative proteins by the bacterial chaperone SecB. Science, 251: 439–443.
402. Randall LL (1992) Peptide binding by chaperone SecB. Science, 257: 241–245.
403. Fekkes P, den Blaauwen T, Driessen AJM (1995) Diffusion-limited interaction between unfolded polypeptides and the Escherichia coli chaperone SecB. Biochemistry, 34: 10078–10085.
404. Glynn M (1985) The Na^+ K^+-Transporting Adenosine Triphosphatase. In: The Enzymes of Biological Membranes, A.Martonosi (eds), vol 3, N–Y: Plenum Press, pp 28–114.
405. Karlish SJD, Yates DW (1978) Tryptophan fluorescence of Na^+-ATPase as a tool for study of the enzyme mechanism. Biochim Biophys Acta, 527: 115–130.
406. Smirnova IN, Lin SH, Faller LD (1995) An equivalent site mechanism for Na^+ and K^+ binging to sodium pump and control of the conformational change reported by fluoresceine 5'-isothiocyanate modification. Biochemistry, 34: 8657–8667.
407. Apell HJ., Roudna M, Corrie JET, Trentham DR (1996) Kinetics of the phosphorylation of Na,K-ATPase by inorganic phosphate detected by a fluorescence method. Biochemistry, 35: 10922–10930.
408. Fedosova NU, Cornelius P, Klodos I (1995) Fluorescent styryl dyes as probes for Na,K-ATPase reaction mechanism: significance of the charge of the hydrophilic moiety of RH dyes. Biochemistry, 34: 16806–16814.
409. Dobretsov GE, Spirin MM, Chekrygin OV, Vladimirov YuA, Kaplun AP, Basharuli VA, Shvets VI (1981) Fluorescent probes, fatty acid derivatives; depth of embedding of chromophore into the lipid layer. Bioorg Chim (in Russian), 7, N 4: 606–612.
410. Spirin MM (1982) Fluorescent Tests to Study the Spatial Structure, Permeability and Surface Charge of Biological Membranes (in Russian). PhD. thesis. Moscow.
411. Cherkasov AS, Bazilevskaya NS (1965) Excited dimers-excimers of anthracene derivatives and concentration quenching of fluorescence. Izv Akad Nauk USSR (in Russian), 29, N 8: 1284–1294.
412. Weber G et al (1971) The use of cholinergic fluorescence probe for study of the re-

ceptor proteolipid. Molec Pharmacol, 7: 530–537.
413. Sudlow G et al (1975) The characterization of two specific drug binding sites on human serum albumin. Molec Pharmacol, 11: 824–832.
414. Dobretsov GE, Liskovcev VV, Vekshin NL (1979) Relationship the structure and affinity to phospholipid membranes with antiarrythmia activity of phenothyasin derivatives. Pharmacol. and Toxicol. (in Russian), N 2: 136–141.
415. Lee JS et al (1994) Rhodamin efflux patterns predict P-glycoprotein substrates in the National Cancer Institute Drug Screen. Molec Pharm, 46: 627–638.
416. Phillips D, Eigenbrot IV, Oldham TC (1996) Spectroscopic studies of drugs used in photodynamic therapy. In: Biomedical Applications of Spectroscopy, vol 25 R.J.H.Clark) (eds), John Wiley, pp 89–141.
417. Denisov YP, Danilov SM (1975) Use of fluorescent probes to study membrane-barbiturates interactions. Biofizika (in Russian), 20, N 6: 1027–1028.
418. Dobretsov GE, Liskovtsev VV, Vekshin NL (1977) Correlation between the antiarrhythmic activity of phenothiazine acyl derivatives and their affinity to phospholipid membranes. Bull Exper Biol i Med (in Russian), N 9: 311–313.
419. Szent-Gyorgyi A (1960) Introduction to a Submolecular Biology, London.
420. Yamazaki T, Ohta N, Yamazaki I, Song PS (1993) Excited-state properties of hypericin: electronic spectra and fluorescence decay kinetics. J Phys Chem, 97: 7870–7875.
421. Azzi A (1975) The application of fluorescent probes in membrane studies. Quart Rev Biophys, 8, N 2: 237–316.
422. Galla HJ, Lulsetti J (1980) Lateral and transversal diffusion and phase transitions in erythrocyte membranes. Biochim Biophys Acta, 596, N 1: 108–117.
423. Guard-Friar D, Chen ChH, Engle AS (1988) Deuterium isotope effect on the stability of molecules: phospholipids. Chem Phys Lipids, 47, N 4: 237–244.
424. Vekshin NL (1985) On membrane viscosity measurements using pyrene excimer. Stud Biophys, 106, N 2: 69–78.
425. Vekshin NL (1987) On using pyrene as a luminescent indicator of viscosity of model and biological membranes. Biologicheskye Nauki (in Russian), N 11: 59–66.
426. Vekshin NL (1988) Localization of fluorescence probes in biomembranes. In: Luminescence Analysis in Medicine and Biology and Appropriate Equipment (in Russian). Riga: RMI, pp 67–68.
427. Blackwell MF, Gounaris K, Barber J (1986) Evidence that pyrene excimer formation in membranes is not diffusion-controlled. Biochim Biophys Acta, 858, N 2: 221–234.
428. L'Heureux GP., Fragata M (1989) Fluorescence characteristics of pyrene and phosphatidylethanolamine-bound pyrene incorporated into lipid vesicles solubilized in media of differing NaCI concentrations. Biophys Chem, 34: 163–168.
429. Hug DH, Hunter JK (1991) Photomodulation of enzymes. J Photochem Photobiol, B: Biol, 10: 3–22.
430. Willner I, Rubin S (1993) Reversible photoregulation of the activities of proteins. Reactive Polymers, 21, N 3: 177–186.
431. Ninnemann H (1991) Participation of the molibdenum cofactor of nitrate reductase from Neurospora crassa in light-promoted condition. J Plant Physiol, 137: 677–682.
432. Lipman RSA, Jorns MS (1996) An unnatural folate stereoisomer is catalytically component in DNA photolyase. Biochemistry, 35: 7968–7973.
433. Kungle AJ, Visser AJWG, Kauffmann HF, Breitenbach M (1994) Time-resolved fluorescence studies of dityrosine in the outer layer of intact yeast ascospores. Bio-

phys .J, 67, N 1: 309–317.
434. Ramsey AJ, Alderfer JL, Jorns MS (1992) Energy transduction during catalysis by *Escherichia coli* DNA photolyase. Biochemistry, 31, N 31: 7134–7142.
435. Le Doan T, Perrouault L, Asseline U, Thuong NT, Rivalle C, Bisagni E, Helene C (1991) Recognition and photo-induced cleavage and cross-linking of nucleic acids by oligonucleotides covalently linked to ellipticine. Antisense Research and Development, 1: 43–54.
436. Honda J, Kandori H, Okada T, Nagamune T, Shichida Y, Sasabe H, Endo I (1994) Spectroscopic observation of the intramolecular electron transfer in the photoactivation processes of nitrile hydratase. Biochemistry, 33, N 12: 3577–3583.
437. Bjorling SC, Zhang CF, Farrens DL, Song PS, Kliger DS (1992) Time-resolved circular dichroism of native oat phytochrome photointermediates J Am Chem Soc, 1992, 114: 4581–4588.
438. Murata Y, Fukutani K (1993) Effects of vibrational excitation in photostimulated desorption. J Electron Spectr Relat Phenomena, 64(5): 533–542.
439. Fain B, Fleurov V, Lin SH (1988) Intermolecular energy transfer in infrared-laser-induced desorption. Chem Phys, 122, N 1: 17–28.
440. Brittain T, Greenwood C, Springall JP, Thomson AJ (1982) The nature of ferrous haem protein complexes prepared by photolysis. Biochim Biophys Acta, 703: 117–128.
441. Chance B, Saronio C, Waring A, Leigh JS (1978) Cytochrome C -cytochrome oxidase interaction at subzero temperatures. Biochim Biophys Acta, 503: 37–55.
442. Lingle R (Ir.), Xu X, Zhu H, Yu S, Hopkins JB, Straub KD (1991) Direct observation of hot vibrations in photoexcited deoxyhemoglobin using picosecond Raman spectroscopy. J Am Chem Soc, 113, N 10: 3992–3994.
443. Petrich J, Martin JL, Houde D, Poyart C, Orszag A (1987) Time-resolved Raman spectroscopy with subpicosecond resolution: vibrational cooling and delocalization of strain energy in photodissociated carbonmonoxyhemoglobine. Biochemistry, 26: 7914–7923.
444. Clarke DL, Collins MA (1992) Simulation of coherent energy transfer in an α-helical peptide by Fermi resonance. Biophys J, 61, N 2: 316–333.
445. Eklund H, Nordstrom B, Zeppezauer E, Soderlund G, Ohlason I, Boiwe T, Sodenberg BO, Tapia O, Branden CL, Akeson A (1976) Three-dimensional structure of horse liver alcohol dehydrogenase at 2.4 Angst resolution J Mo. Biol, 102, N 1: 27–59.
446. Eklund I, Samama JP, Wallen L, Branden CI., Akeson A, Jones TA (1981) Structure of triclinic ternary complex of horse liver alcohol dehydrogenase at 2.9 Angst. resolution. J Mol Biol, 147: 561–587.
447. Theorell H, Tatemoto K (1971) Excitation, transfer in complexes of horse liver alcohol dehydrogenase. Arch Bioch Brophys, 142, N 1: 69–82.
448. Lehninger A (1985) The Principles of Biochemistry (in Russian), vol. 1–3. Moscow: Mir.
449. Vekshin NL (1992) Photophysical processes in the NADH-alcohol dehydrogenase complex. J Photochem Photobiol, B: Biol, 12: 295–303.
450. Ataullakhanov FI, Zhabotinsky AM (1975) Photoinduced reduction of ferriperoxidase; the reaction with reduced nicotinamide adenine dinucleotide. Biofizika (in Russian), 20, N 4: 596–601.
451. Cilento G (1980) Photobiochemistry in the dark. Photochem Photobiol Rev, 5: 199–228.

452. Haun M, Duran N, Gilento G (1978) Energy transfer from enzymically generated triplet carbonyl compounds to the fluorescent state of flavins. Biochem Biophys Res Commun, 81: 779–784.
453. Kachar B, Zinner K, Vidigal CCC, Shimizu Y, Cilento G (1979) Excitation of eosin when catalyzing electron transport in biochemical systems. Arch Biochem Biophys, 195: 245–247.
454. Baskakov IV, Voeikov VL (1996) The role of electron excited species in biochemical processes. Biokhimiya (in Russian), 61, N 7: 1169–1181.
455. Molecular Mechanisms of the Biological Action of Optical Radiation (1988), A.B.Rubin (eds). Moscow: Nauka, (in Russian).
456. Salet C, Moreno G, Atlante A, Passarella S (1991) Photosensitization of isolated mitochondria by hematoporphyrin derivative (photofrin) effects on bioenergetics. Photochem Photobiol, 53, N 3: 391–393.
457. Ludkovskaya RG, Burmistrov YuM (1971) Photobioelectrical processes in exciting cells. In: Biophysics of Living Cell, G.M.Frank (eds), vol 2 (in Russian). Pushchino: Acad .Sci, pp 50–67.
458. Ninnemann H, Strasser RJ, Butler WL (1977) The superoxide anion as electron donor to the mitochondrial electron transport chain. Photochem Photobiol, 26, N 1: 41–47.
459. Belyanovich LM, Nikol'skayaVP, Rudenok AN, Konev SV (1996) Spectral characteristics of cyanid-inhibited cytochrome c oxidase in its photoreactivation and interaction with ATP. Zh Prikl Spektr (in Russian), 63, N 3: 395–403.
460. Racker E, Stoeckenius W (1974) Reconstitution of purple membrane vesicles catalyzing light-driven proton uptake and ATP formation. J Biol Chem, 249: 662–663.
461. Ryrie YJ, Gritchley Gh, Tillberg JE (1979) Structure and energy-linked activities in reconstituted bacteriorhodopsin-yeast ATPase proteoliposomes. Arch Bibchem Biophys, 198, N 1: 182–194.
462. Scheuerlein R, Braslavsky SE (1987) Induction of chloroplast movement in the alga Mougeotia by polarized nanosecond dye-laser pulses. Photochem Photobiol, 46, N 4: 525–530.
463. Karu T, Smolyanmova N, Zelenin A (1991) Long-term and short-term responses of human lymphocytes to He-Ne laser radiation. In: Lascrs in the Life Sciences 4 (3), Harwood Acad Press Pub GmbH, pp 167–178.
464. Goldman YE, Hibberd MG, McCray JA, Trentham D R (1982) Relaxation of muscle fibres by photolysis of caged ATP. Nature, 300, N 5894: 701–705.
465. Peckham M, Ferenczi MA., Irving M (1994) A birefringence study of changes in myosin orientation during relaxation of skinned muscle fibers induced by photolytic ATP release. Biophys J, 67: 1141–1148.
466. Vekshin NL, Mironov GP (1982) Flavin-dependent consumption of oxygen in mitochondria upon illumination. Biofizika (in Russian), 27, N 3: 537–539.
467. Konev SV, Rudenok AN (1992) Effects of cross-linking reagents on the respiratory chain of mitochondria. Biofizika (in Russian), 37, N 5: 939–941.
468. Radda GK, Calvin M. Chemical and photochemical reductions of flavin nucleotides and analogs. Biochemistry, 1964, 3, 384–389.
469. Vekshin NL, Mironov GP (1981) Oxidation of NADH by singlet oxygen formed with participation of triplet flavin. Biofizika (in Russian), 26, N 6: 953–959.
470. Vernon LP (1959) Photochemical oxidation and reduction reactions catalyzed by flavin nucleotides. Biochim Biophys Acta, 36, N 1: 177–185.

471. Krasnovsky AA, Brin GP, Nikandrov VV (1975) Light excitation of reduced pyrimidine nucleotides: transfer of electrons to ferrodoxin and methyl viologeti. Dokl Akad Nauk USSR (in Russian), 220, N 5: 1214–1219.
472. Nikandrov VV, Bnin GP, Krasnovsky AA (1978) Light activation of NADH and NADFH. Biokhimiya (in Russian), 43, N 4: 636–645.
473. Byteva IM, Gurinovich GP, Petsold OM (1975) On the mechanism of photooxidation of porphyrines by oxygen. Biofizika (in Russian), 20, N 1: 51–55.
474. Timinsky YV, Sinyakov GN, Byteva IM, Gurinovich GP (1978) The mechanism of photoregeneration of reduced porphyrine in the presence of oxygen. Biofizika (in Russian), 23, N 3: 411–413.
475. Krasnovsky AA, Ghernysheva EK, Kritsky MS (1987) Study of the role of active forms of oxygen in the flavin-sensitized oxidation of NADH. Biokhimiya (in Russian), 52, N 9: 1474–1483.
476. Peters G, Rodgers MAJ (1981) Single-electron transfer from NADH analogoues to singlet oxygen. Biochim Biophys Acta, 637, N 1: 43–52.
477. Barsky EL, Kamilova FD, Remennikov VG, Samuilov VD (1986) The inhibiting effect of azide on the photoreduction of oxygen by chlorophyll A in micelles of triton x-100. Biofizika (in Russian), 31, N 5: 789–792.
478. Fritz BJ, Ninnenmann H (1985) Photoreactivation by triplet flavin and photoinactivation by singlet oxygen of *Neurospora crassa* nitrate reductase. Photochem Photobiol, 41, N 1: 39–45.
479. Straub KD (1974) A solid state theory of oxidative phosphorylation. J Theor Biol, 44, N 2: 191–206.
480. Aggarwal BB, Quintanilha AT, Cammack R, Packer L (1978) Damage to mitochondrial electron transport and energy coupling by visible light. Biochim Biophys Acta, 502: 367–382.
481. Vekshin NL (1983) Photoactivation of electron and energy transfer in biochemical «dark» systems. In: Physical-chemical Basis of Cell Function (in Russian). Pushchino: Acad Sci USSR, pp 82–86.
482. Vekshin NL (1991) Light-dependent ATP synthesis in mitochondria. Biochem Intern, 25, N 4: 603–611.
483. Salet C, Passarella S, Quagliariello E (1987) Effects of selective irradiation on mammalian mitochondria. Photochem Photobiol, 45, N 3: 433–438.
484. Mosolova IM, Gorskaya IA, Sholts KF, Kotelnikova AV (1975) Isolation of intact mitochondria from rat liver. In: Methods of the Present-Day Biochemistry (in Russian). Moscow: Nauka, pp 45–47.
485. Nedelina OS, Brzhevskaya ON, Lozinova TA, Piskunov MA, Kayushin LP (1983) Effect of the visible light on the ATP-synthesizing activity of rat liver mitochondria. Biofizika (in Russian), 28, N 1: 341–342.
486. Passarella S, Casamassima E, Molinari S, Pastore D, Quaglianello E., Catalano I.M., Cingolani A (1984) Increase of proton electrochemical potential and ATP synthesis in rat liver mitochondria irradiated in Vitro by helium-neon laser. FEBS Lett, 175, N 1: 95–99.
487. Dmitriev LF, Ivanova MV, Ivanov II (1990) Liver mitochondrial ATP synthesis can be inhibited and promoted by UV generation of superoxide. Biologicheskie Membrani (in Russian), 7, N 9: 961–965.
488. Stitt M, Lilley R. McC, Heldt HW (1982) Adenine nucleotide levels in the cytosol, chloroplasts, and mitochondria of wheat leaf protoplasts. Plant Physiol, 70: 971–977.

489. Takabe T, Hammes GG (1981) pH dependence of ATP synthesis and hydrolysis catalyzer by reconstituted chloroplast coupling factor. Biochemistry, 20, N 24: 6859–6864.
490. Pitard B, Richard P, Dunach M, Rigaud JL (1996) ATP synthesis by the F_0F_1 ATP synthase from thermophilic *Bacillus* PS3 reconstituted into liposomes with bactenorhodopsin Eur J Biochem, 235: 777–788.
491. Henry ER, Baton WA, Hochstrasser RM (1987) Molecular dynamics simulation of cooling in laser-excited hem-proteins. Biophys J, 51, N 2, Pt 2, 404a.
492. Genberg L, Heisel F, McLendon G, Miller JD (1988) Vibrational energy relaxation processes in heme proteins. J Luminescence, 40–41: 571–572.
493. Vekshin NL (1990) Thermal hypothesis of coupling ATP synthesis to electron transfer in biological membranes. Comm. Molec. Cell. Biophys., 7, N 1: 17–25.
494. Grubmeyer C, Cross RL, Penefsky HS (1982) Mechanism of ATP hydrolysis. Rate constants for elementary steps in catalysis at a single site. J Biol Chem, 257: 12092–12100.
495. Feldman RI, Sigman DS (1982) The synthesis of enzyme-bound ATP by soluble chloroplast coupling factor. J Biol Chem, 257, N 4: 1676–1683.
496. Sakamoto J, Tonomura Y (1983) Synthesis of enzyme-bound ATP by mitochondrial soluble Fl-ATPase in the presence of DMSO. J Biochem, 93: 1601–1614.
497. Muller AW (1985) Thermosynthesis by biomembranes: energy gain from cyclic temperature changes. J Theor Biol, 115, N 3: 429–453.
498. Muller A.W (1983) Thermoelectric energy conversion could be an energy source of living organisms. Phys Lett, 96a: 319–321.
499. Nurmukhametov RN (1971) Absorption and Luminescence of Organic Compounds (in Russian). Moscow: Chemistry.
500. Harris DA, Bashford CL (1987) Spectrophotometry and Spectrofluorimetry. A Practical Approach. Oxford: IRL Press.
501. Batzri S, Korn ED (1973) Single bilayer liposomes prepared without sonication. Biochim Biophys Acta, 298, N 4: 1015–1019.
502. Kirby EP, Steiner RF (1970) The tryptophan microenvironments in apomyoglobin. J Biol Chem, 245: 6300–6305.
503. Karalis VN, Korneeva EA (1970) Apparatus for Fluorescence Analysis (in Russian). Moscow: Publisher of standards.
504. Afonin EI, Lee ME (1981) Author's certificate No 842511. Bull Izobr (in Russian), N 24: 152–153.
505. Muller G, Schaldach M (1981) Patent BRD N 2937352, cl. G01 N 21/01.
506. Kochetkov NK, Budovsky EI, Sverdlov ED, Simukova NA, Turchinsky MF, Shibaev VN (1970) Organic Chemistry of Nucleic Acids (in Russian). Moscow: Chemistry.
507. Dolinnaya NG, Gromova ES (1983) Complementary interactions of oligonucleotides. Uspekhi Khimii (in Russian), 52, N 1: 138–167.
508. Norden B (1978) Applications of linear dichroism spectroscopy. Appl Spectr Rev, 14: 157–248.
509. Litvinov IS, Obraztsov VV (1982) Study of the viscosity of free and protein-bound lipids in membranes. Biofizika (in Russian), 27, N 1: 81–86
510. Vekshin NL (1983) Application of the nonradiative transfer of electronic excitation energy for studying the spatial organization of biomembraneous protein-lipid complexes (in Russian). Preprint. Pushchino: Acad Sci USSR.
511. Anderson SR, Brunori M, Weber G (1970) Fluorescence studies of Aplysia and

sperm whale apomyoglobins. Biochemistry, 9, N 24: 4723–4729.
512. Frisell WR, Mackenzie CG (1959) The photochemical oxidation of DPNHh with riboflavin phosphated. Proc Natl Acad Sci, 45: 1568–1575.
513. Russian Federation Patent No 1412452, class G01N21/03, 1984, 1993.
514. Russian Federation Patent No 1254357, class G01N21/03, 1984, 1993.
515. Marriott G (1994) Caged protein conjugates and light-directed generation of protein-activity: preparation, photoactivation, and spectroscopic characterization of caged g-actin conjugates. Biochemistry, 33, N 31: 9092–9097.
516. Ding J, Starling AP, East JM, Lee AG (1994) Binding-sites for cholesterol on Ca^{2+}-ATPase studied by using a cholesterol-containing phospholipid. Biochemistry, 33, N 16: 4974–4979.
517. Hutchinson J, Noe LJ (1984) A study of the photodissociation of the CO and O_2 forms of hemoglobin and myoglobin by picosecond absorption spectroscopy. IEEE J Quant Electron, QE-20: 1353–1362.
518. Sutherland BM, Sutherland JC (1969) Inhibition of pyrimidine dimer formation in DNA by cationic molecules: role of energy transfer. Biophys J, 91045–1055.
519. Duhamel J, Kanyo J, Dinter-Gottlieb G, Lu P (1996) Fluorescence emission of ethidium bromide intercalated in defined DNA duplexes: Evaluation of hydrodynamics components. Biochemistry, 35: 16687–16697.
520. Parkhurst KM, Brenowitz M, Parkhtirst LJ (1996) Simultaneous binding and bending of promoter DNA by the TATA binding protein. Biochemistry, 35: 7459–7465.
521. Maliwal BP, Kusba J, Lakowicz JR (1995) Fluorescence energy transfer in one dimension: frequency-domain fluorescence study of DNA-fluorophore complexes. Biopolymers, 35: 245–255.
522. Mergny JL, Boutorine AS, Garestier T, Belloc F, Rougee M, Bulychev NV, Koshkin AA, Bourson J. Lebedev AV, Valeur B, Thuong NT, Helene C (1994) Fluorescence energy transfer as a probe for nucleic acid structures and sequences. Nucleic Acids Research, 22, N 6: 920–928.
523. Ronzina OA, Kravchenko EB (1988) Influence of light on the adenine nucieotide system in the barley mitochondria, In: Questions on the correlation between photosynthesis and respiration, V.L. Voznesenskii (eds). Tomsk: TGU (in Russian), 48–50.
524. Belousov AV, Kovarsky VA, Merlin ET, Yastrebov BS (1993) Enzyme reaction in the external electromagnetic field Biofizika (in Russian), 38, N 4: 619–626.
525. Vekshin NL (1998) Protein sizes and stoichiometry in the chaperone SecB-RBPTI complex estimated by ANS fluorescence. Biochemistry (Moscow), vol 63, N 4: 485–488.
526. Weigand R, Rotemund F, Penzkofer A (1997) Aggregation Dependent Absorption Reduction of Indocyanine Green. J Phys Chem, 101: 7729–7734.
527. Gryczynski Z, Bucci E (1993) A new frront-face optical cell for measuring weak fluorescent emission with time resolution in the picosecond time scale. Biophysical Chemistry, 48: 31–38.
528. Gruebelle M, Sabelko J, Ballew R, Ervin J (1998) Laser temperature Jump Induced Protein Refolding. Acc Chem Res 31: 699–707.
529. McMahon LP, Yu HT, Vela MA, Morales GA, Shui L, Fronczek FR, McLaughlin ML, Barkley MD (1997) Confomer Interconversion in the State of Constraind Tryptophan Derivates. J Phys Chem, 101: 3269–3280.
530. Toptygin D, Savtchenko R, Meadow N, Brand L (2001) Homogeneous Specrtally- and Time-Resolved Fluorescence Emission from Single-Tryptophan Mtants of IIA^{Glc}

Protein. J Phys Chem, 105: 2043–2055.
531. Vincent M, Gilles AM, de la Sierra IML, Briozzo P, Barzu O, Gallay J (2000) Nanosecond Fluorescence Dynamic Stokes of Trypophan in a Protein Matrix. J Phys Chem, 104: 11286–11295.
532. Vivian JT, Callis PR (2001) Mechanisms of Tryptophan Fluorescence Shifts in Proteins. Biophysical Journal, vol 80: 2093–2109.
533. Gryczynski Z, Lubkowski J, Bucci E (1995) Heme-Protein Interactions in Horse Heart Myoglobin at Neutral pH and Exposed to Acid Investigated by Time-resolved Fluorescence in the Pico- to Nanosecond Time Range. The Journal of Biological Chemistry, vol 270, No 33: 19232–19237.
534. van den Berg PAW, van Hoek A, Walentas CD, Perham RN, Visser AGWG (1998) Flavin Fluorescence Dynamics and Photoinduced Electron Transfer in *Escherichia coli* Glutathione reductase. Biophysical Journal, vol 74: 2046–2058.
535. Lomtev AS, Sharova IV, Vekshin NL (2001) Flavin and Ubiquinone of Mitochondrial NADH Dehydrogenase Are not Involved in the Electron Transfer to Artifical Acceptors. Doklady Biochemistry and Biophysics, vol 376: 1–3.
536. Sukharev VI, Vekshin NL (2000) The Globule Diameter and Electron nd Conformational properties of the Flavoprotein Fragment of Mitochondrial NADH Dehydrogenase Studied by Fluorescence Spectroscopy. Russian Journal of Bioorganic Chemistry, vol 26, No 10: 651–655.

Index

Acceptor 1, 5, 44–47, 51, 74, 84, 90–94, 100–110, 128, 132, 153
Aggregate 8, 11, 13, 14, 17–23, 30–35, 44, 48
Alcohol dehydrogenase 72, 148, 174–179
ANS 52, 53, 151–158, 161–164
Anthracene 8, 32–35, 95, 96
Aromatic hydrocarbon 32, 34, 84, 94, 96, 98, 100
ATP 151, 158–159, 163–164, 179, 184, 193–199

Bovine serum albumin 36, 58–59, 66, 176, 178

Chaperon 155, 157, 163–164
Chloroplast 12, 31, 32, 45, 149
Cluster 12, 32–34
Conformational change 5, 61, 72, 107, 128, 132, 145, 172–173
Cuvette 45–49
  mirror 53
  multipass 52–55

Diffusion 48, 150, 152, 153, 160–162, 166–168
DNA 13–16, 23–26, 41, 153, 173
Donor 5, 41, 45–49, 90–110, 129, 133, 139

Effect
  inner filter 41
  sieve 13, 14, 17
Erythrocyte 30, 31, 33, 44, 51, 72
Ethidium bromide 26, 111, 116, 153
Excimer 1, 3, 5, 90–100, 165–171
Exciplex 1, 5, 84, 85, 90–100
Excited state 4, 5, 57, 65, 72, 74, 77–83, 94, 102, 105, 107

Extinction coefficient 4, 7–10, 16, 27–32, 44–46

Flavin 10, 88, 189–192
Fluorescence pharmacology 162
Fluorescent probe 58, 119, 151–164

Glycerol 56, 57, 63–68, 74–81, 88

Hypochromism 11, 12, 14, 15–17, 18–35, 42, 116, 117

Indole 69–72, 74–88, 90, 91, 100
Inductive resonance 102, 105, 110
Internal conversion 1, 3, 71, 92, 110
Intersystem crossing 1, 3, 4

Lambert–Beer law 9
Lifetime 4, 5, 48, 57, 62–73, 75, 88, 90–93, 99
Light absorption 1, 2, 3, 7, 20

Microreabsorption 41–45, 51
Microscreening 41–45, 51
Migration 16, 62, 67, 107–110, 117–121, 125–127, 150
Mitochondria 37–40, 143, 147–150, 174, 179–199

Na^+K^+-ATPase 158, 159
NADH 10, 62, 143, 147, 148, 150, 173, 174–183, 187–192
  dehydrogenase 173, 179–183
NATA 69, 75–87
Nucleotide 16, 23, 24, 26, 35

Oligonucleotide 13, 16, 19, 23–26, 173
Oxygen 4, 7, 27, 73, 90, 185–192

Parvalbumin 86, 87

Phosphorescence 4, 5, 10, 71, 72, 90
Phosphorylation 193–199
Photoactivation 172, 184–191
Photobleaching 10
Photodesorption 173, 177, 179
Photolysis 71, 173, 179–182
Polarity 78, 91, 97
Polarization 4, 16, 31, 49, 62, 63, 73
Polyadenilic acid 115
Probability of absorption 7, 16-20, 23
Pyrene 142–147, 152, 162, 166–171

Quantum yield 5, 10, 42, 45, 63, 65, 68, 71, 94, 98, 105, 112–115
Quenching 10, 48, 57, 62, 72, 78–88, 91–100, 105, 110, 115, 119, 123–141

Reabsorption 48–51, 102, 106, 108
Rhodamine 47–51

Sarcoplasmic reticulum 37, 40, 66
Scattering 4, 10, 13, 14, 17, 23, 27–30, 36–38, 69
Screening 10, 17–35, 42–51, 102, 106, 108

Singlet oxygen 187–192
Stacking 12, 15, 31
Suspension 13, 17, 37–45, 51
Synchronous scanning 58, 61

Thermal coupling 197, 199
Time-resolved spectroscopy 57, 75, 76
Transfer
   electron 3
   energy 10, 45, 48, 49, 110–142
   fractional energy 79
   hot energy 109, 110, 121
Tryptophan 2, 9, 10, 27, 28, 35, 56–61, 74, 88, 89, 94, 99, 122–141, 166, 173, 178
Tyrosine 9, 10, 27, 28, 35, 56–61, 84, 122–141, 173

Vibrational relaxation 68, 97, 101, 105, 108–110, 127, 177, 179
Vibronic peak 170
Viscosity 64, 78, 83, 84, 107, 108, 166–171
Volume reabsorption 45–51, 136

Printing (Computer to Film): Saladruck Berlin
Binding: Stürtz AG, Würzburg